AN INTRODUCTION TO THE NEW WORLD ORDER

GEORGE KLENOWSKI

FOR THE ETERNAL SURVIVAL OF THE HUMAN RACE WITH DIGNITY

Copyright © 2016 by George Klenowski

AN INTRODUCTION TO THE NEW WORLD ORDER
BY GEORGE KLENOWSKI

Publishing data:

ISBN : 1537245023

Printed by:

CreateSpace 2016

Kindle eBook 2016

All rights reserved. No part of this publication may be reproduced, stored in a retrieval system or transmitted in any form or by any means, electronic, mechanical, photocopying, recording or otherwise, without prior written permission of the copyright owners and publishers.

AN INTRODUCTION TO THE NEW WORLD ORDER

CONTENTS

PAGE

1.0	INTRODUCTION		1
2.0	MATTER, SPACE AND TIME		4
	2.1	Origins of Modern Physics	4
		2.1.1 Sir Isaac Newton (1643 to 1727)	4
		2.1.2 Thermodynamics	6
		2.1.3 Albert Einstein (1879 to 1955)	7
		2.1.4 Quantum Mechanics	10
	2.2	Matter	13
	2.3	Space	25
	2.4	Time	26
3.0	THE UNIVERSE		32
	3.1	General	32
	3.2	Astronomy	36
	3.3	Interstellar Medium	37
	3.4	Molecular Cloud	38
	3.5	Nebula	38
	3.6	Protostar	39
	3.7	Stars	39
		3.7.1 Definition	39
		3.7.2 Star Classification	40
		3.7.3 Main Sequence Star	41
		3.7.4 Post-Main Sequence Events	42
		3.7.5 Neutron Star	43
		3.7.6 Pulsar and Magnetar	44

		3.7.7	Binary Stars 44

- 3.7.7 Binary Stars 44
- 3.7.8 Constellation 44
- 3.7.9 Galaxy 44
- 3.7.10 Quasar 46
- 3.7.11 Globular Cluster 46
- 3.7.12 Galaxy Groups, Clusters and Superclusters 46

3.8 The Solar System 47
- 3.8.1 The Sun 49
- 3.8.2 Mercury 51
- 3.8.3 Venus 51
- 3.8.4 Earth 52
- 3.8.5 Earth's Moon 53
- 3.8.6 Mars 55
- 3.8.7 Asteroid Belt 56
- 3.8.8 Jupiter 56
- 3.8.9 Saturn 58
- 3.8.10 Uranus 59
- 3.8.11 Neptune 59
- 3.8.12 Trans-Neptunian Region 60
  - 3.8.12.1 Kuiper Belt 60
  - 3.8.12.2 Scattered Disc 61
  - 3.8.12.3 Heliosphere 61
  - 3.8.12.4 Oort Cloud 62

3.9 Implications for Space Travel 63

4.0 THE EVOLUTION OF LIFE ON EARTH 72
- 4.1 Introduction 72
- 4.2 Features of Evolution 72
- 4.3 Geological Time Scale 74
- 4.4 Hadean 74
- 4.5 Archean 78

- 4.6 Proterozoic 81
  - 4.6.1 Introduction 81
  - 4.6.2 Great Oxygenation Event 82
  - 4.6.3 Huronian Galaciation 84
  - 4.6.4 Columbia Supercontinent 84
  - 4.6.5 Rodinia Supercontinent 84
  - 4.6.6 Snowball Earth 85
  - 4.6.7 Proterozoic Life 85
- 4.7 Paleozoic 87
  - 4.7.1 Introduction 87
  - 4.7.2 Cambrian 88
  - 4.7.3 Ordovician 90
  - 4.7.4 Silurian 91
  - 4.7.5 Devonian 92
  - 4.7.6 Carboniferous 94
  - 4.7.7 Permian 96
- 4.8 Mesozoic 98
  - 4.8.1 Introduction 98
  - 4.8.2 Triassic 98
  - 4.8.3 Jurassic 101
  - 4.8.4 Cretaceous 102
- 4.9 Cenozoic 105
  - 4.9.1 Introduction 105
  - 4.9.2 Paleocene 105
  - 4.9.3 Eocene 109
  - 4.9.4 Oligocene 111
  - 4.9.5 Miocene 113
  - 4.9.6 Pliocene 115
  - 4.9.7 Pleistocene 117
  - 4.9.8 Holocene 123
- 4.10 Human Evolution 126

4.11 Extinction Events and Implications for the Human Race 135

    4.11.1 Introduction 135

    4.11.2 Effects of Extinction Events on the Human Race 137

5.0 THE NEW WORLD ORDER 140

5.1 Introduction 140

5.2 Human Population 141

5.3 Genetic Engineering and Cloning 143

5.4 Society and Infrastructure 147

5.5 Technological Advances Which Will Assist Human Evolution 149

    5.5.1 Goals in Genetic Engineering 150

    5.5.2 Protection of Humans from Cataclysmic Events 151

    5.5.3 Future Trends in Science, Engineering, Technology and Recreation 151

    5.5.4 Agriculture and Chemosynthesis 153

    5.5.5 Solar Energy 157

    5.5.6 Nuclear Fusion 158

    5.5.7 Superconductivity 162

    5.5.8 Nanotechnology 164

    5.5.9 Teleportation 166

    5.5.10 Space Travel and Extraterrestrial Colonisation 168

    5.5.10.1 Introduction 168

    5.5.10.2 Space Travel 168

    5.5.10.3 Human Adaptation to Space Travel ... 175

    5.5.10.3.1 Vacuum Suit 175

    5.5.10.3.2 Vacuum Chamber 175

    5.5.10.3.3 Artificial Gravity 176

    5.5.10.3.4 Suspended Animation 177

    5.5.10.3.5 Genetic Engineering 177

5.5.10.4 Extraterrestrial Colonisation 178
6.0 CONCLUSION 182
7.0 REFERENCES 183

TABLES

Table 1.	Classes of Particles 18	
Table 2.	The Standard Model of Elementary Particles (after Reference 1) 19	
Table 3.	Distances of Solar System Bodies from Earth 63	
Table 4.	General Geological Time Scale 75	
Table 5.	Hadean Geological Time Scale 76	
Table 6.	Archean Geological Time Scale 78	
Table 7.	Proterozoic Geological Time Scale 83	
Table 8.	Paleozoic Geological Time Scale 89	
Table 9.	Mesozoic Geological Time Scale 99	
Table 10.	Cenozoic Geological Time Scale 106	
Table 11.	Chronology of Quaternary Glacial Cycles 119	
Table 12.	World Population Trend 142	

FIGURES

Figure 1.	Universal Time Number Line 30	
Figure 2.	A Traveller in Relative Time Related to the Universal Time Number Line 31	
Figure 3.	Hertzsprung-Russell Diagram (after Reference 2) 40	

v

1.0 INTRODUCTION

The human race has reached the evolutionary crossroads and has the choice of indefinite, dignified survival by adopting the New World Order or it can suffer an ignominious extinction. The World now has a human population of 7.4 billion most of whom are non-productive and rampantly consumeristic. The gross overpopulation is destroying the environment, exterminating Earth's biodiversity in a mass extinction event and rapidly depleting natural resources. In 1900 Earth's population was 1.6 billion. By 2011 the population had increased to 7 billion and current projection is 10 billion people by 2050. This is unsustainable and will result in global starvation, particularly if there is a mass extinction event such as an exploding supervolcano or a major meteor impact on Earth. In about 5 billion years the Sun will incinerate the Earth when it becomes a red giant.

Worldwide society is dysfunctional with a morass of unnecessary bureaucracy and is regressing through a lack of ethics and proper human values. Misguided sociological trends are detracting from the true reason for human existence, which is permanent, dignified, technological survival in a physically hostile Universe. Rampant, uncontrolled breeding is rapidly degrading the human gene pool. The New World Order is urgently needed.

For a person to comprehend the tenuousness of life and the enormous effort required for indefinite human survival a knowledge of science and history is necessary. To progress to the New World Order a thorough understanding of matter, space, time and the Universe is

AN INTRODUCTION TO THE NEW WORLD ORDER

a prerequisite. The human race cannot survive without space travel and extraterrestrial colonisation.

Throughout Earth's history over the last 4.6 billion years a number of mass extinction events have occurred, decimating much life. Such events have been caused by large meteor impacts, exploding supervolcanoes and gamma-ray bursts which affect the Earth's atmosphere, including the ozone layer. Mass extinctions are well documented and will occur again. An extinction event is currently occurring due to human overpopulation. The human race needs to be prepared for future mass extinction events which will affect climate and human survival. During its travel in the Milky Way Galaxy the Earth could intercept galactic clouds, newly formed stars or black holes which would obliterate the planet.

A number of changes are required in the New World Order. The human population needs to be reduced to a maximum of 2 billion people. Even such a population would be difficult to maintain in a volcanic winter. To be prepared for mass extinction events adequate food storages are required as well as underground shelters.

Human evolution needs to be systematically accelerated so that rapid technological advances required for human survival are possible. Urgent research is required into genetic engineering, human cloning and the development of embryo farms producing genetically diverse offspring with a range of mental and physical capabilities.

Globalisation is necessary for rational, sociological advances. An efficient education system is required for rapid future progress in science, engineering and technology which are essential for human survival in a hostile Universe. With a sensible, genetically engineered, globalised population the New World Order will be unshackled from the current bureaucratic quagmire and warped human values. Rapid progress

1.0 INTRODUCTION

can then occur in technological advancement and the necessary requirement for successful space travel and extraterrestrial colonisation.

Exciting research fields which will be available in the New World Order include genetic engineering and human cloning, development of techniques for the protection of humans from cataclysmic natural events, advances in agriculture and chemosynthesis, solar energy, nuclear fusion, superconductivity, nanotechnology teleportation and space travel and extraterrestrial colonisation.

With the adoption of the New World Order, which would be accompanied by rapid evolutionary and technological advances, the human race will be assured of permanent survival.

2.0 MATTER, SPACE AND TIME

A fundamental understanding of the concepts of matter, space and time is required for the human mind to comprehend the physical reality of the Universe. Without physics, chemistry and mathematics there would be no scientific or technological advancement on Earth. Rapid progress in science and technology is the only way that the human race can avoid extinction.

2.1 Origins of Modern Physics

2.1.1 Sir Isaac Newton (1643 to 1727)

Sir Isaac Newton is probably the greatest scientist in human history. His achievements included describing universal gravitation and the three laws of motion, elucidating the principles of conservation of momentum and angular momentum, and developing differential and integral calculus at the same time as Gottfried Leibniz.

Absolute theory of space and time was propounded by Isaac Newton. In this theory space and time are real things, being containers of infinite extension or duration within which the whole succession of natural events in the World has a definite position. Things may be at rest or moving and this is not simply a matter of their relations to other objects changing. Newtonian space is an emptiness in which events occur without being influenced by space. Such space is known as flat. Newtonian space on a cosmic scale goes on forever.

Newton's three laws of motion are described below.

1. An object continues in a state of rest or motion at a constant speed along a straight line unless acted

upon by an external force. This law is sometimes called the law of inertia. Inertia is the natural tendency of an object to remain at rest or in motion at a constant speed along a straight line. The mass of an object is a quantitative measure of inertia.

2. When an external force acts on an object of given mass, the acceleration that results is directly proportional to the net force and has a magnitude that is inversely proportional to the mass. The direction of acceleration is the same as the direction of the net force.

3. Whenever one body exerts a force on a second body, the second body exerts an opposite directed force of equal magnitude on the first body (i.e. for every action there is an equal but opposite reaction).

Newton's law of gravitational attraction states that every particle in the Universe exerts an attractive force on every other particle. A particle is a piece of matter small enough to be regarded as a mathematical point. The attractive force is proportional to the mass of each particle. The further apart the particles, the smaller is the force.

Newton's law indicates that there is no unique standard of rest which means it is not possible to determine whether two events which occurred at different times, occurred in the same position in space. The non-existence of absolute rest means that an event cannot be given an absolute position in space.

Newton believed in absolute time, meaning that the interval of time between two events could be accurately and unambiguously measured. Time was unrelated to space. These apparently rational views appear to fail as light speed is approached.

2.1.2 Thermodynamics

Thermodynamics are based on fundamental laws of heat and work. Sadi Garnot, who is known as the father of thermodynamics, published *"Reflections on the Motive Power of Fire"* in 1824. The term thermodynamics was introduced by James Joules in 1849 to describe the science of relations between heat and power. The object of focus is the system and the environment in the surroundings. A system and its surroundings can be separated by a diathermal wall which permits the flow of heat or an adiabatic wall which is a perfect insulator.

In thermodynamics there are four laws which are defined below.

1. The zeroth law of thermodynamics states that two systems individually in thermal equilibrium with a third system are in thermal equilibrium with each other. This law establishes temperature as the indicator of thermal equilibrium.

2. The first law of thermodynamics is the conservation of energy principle applied to heat, work and the change in internal energy. The internal energy of a system changes from an initial value to a final value due to heat and work.

3. The second law of thermodynamics is known as the heat flow statement. Heat flows spontaneously from a system at a higher temperature to a system at a lower temperature and does not spontaneously flow in the reverse direction. Entropy is a measure of the unavailable energy in a thermodynamic system and can be regarded as the amount of disorder in a system. Reversible processes do not change the total entropy of the Universe, but irreversible processes increase the entropy of the Universe. The second law of thermodynamics, stated in terms of entropy, is that the total entropy of the Universe does not change

when a reversible process occurs but increases when an irreversible process occurs.

4. The third law of thermodynamics states that it is not possible to lower the temperature of any system to absolute zero in a finite number of steps.

The laws of thermodynamics and in particular entropy are important in understanding the behaviour of matter.

2.1.3 *Albert Einstein (1879 to 1955)*

Albert Einstein is known as the father of modern physics. He propounded the special theory of relativity in 1905. The theory is designated special because it applies the principle of relativity only to frames in uniform relative motion. In this theory an event is a physical occurrence at a specific place and time. A reference frame is a set of three-dimensional points (a co-ordinate system) and a clock. An inertial reference frame is one in which Newton's law of inertia is valid. Rotating and accelerating reference frames are not inertial reference frames. Einstein's theory of relativity is based on two fundamental assumptions.

1. The Principle of Relativity. The laws of physics are the same in all inertial reference frames.

2. The Constancy of the Speed of Light. The speed of light in a vacuum, measured in any inertial reference frame has the same value of c (3.00×10^8 m/sec) irrespective of how fast the source of light and the observer are moving relative to each other.

Because all the laws of physics are the same for all inertial reference frames there is no experiment that distinguishes between an inertial frame that is at rest and one that is moving at a constant velocity. Since there is no preferred inertial system it is impossible to detect absolute motion. The term absolute velocity is therefore meaningless. Einstein postulated that only relative velocity between objects (not their absolute velocities) can be measured and is physically meaningful.

Einstein's second postulate obviates the need for an ether and states that light always moves at speed c (3.00×10^8 m/sec) with respect to any inertial observer. Although defying logic, this property of light has been experimentally verified many times.

The classical concepts of time, length, momentum, kinetic energy and the addition of velocities developed by Newton and other scientists are not incorrect, but are only limited to speeds that a very small when compared to the speed of light. Relativistic concepts apply to and are consistent at all speeds from zero to light speed.

Special relativity has some startling consequences. The distance between two points and the time interval between two events are dependent on the reference frame in which they are measured. Absolute length and absolute time cannot occur in relativity. Events at different locations which occur simultaneously in one reference frame are not simultaneous in another frame moving uniformly past the first. Results from this theory include length contraction, time dilation and relativity of simultaneity which contradicts a fundamental premise of Newtonian mechanics that the duration of the time interval between two events is equal for all observers. Einstein abandoned the assumption of simultaneity, accepting rather that a time interval measurement is dependent on the reference frame in which the measurement is made. Because there is no preferred inertial frame of reference, there is no correct answer for time events viewed from two reference frames.

The two postulates of special relativity demonstrate the interchangeability of matter and energy in Einstein's mass-energy equivalence formula $E = mc^2$, where energy is equal to mass times the speed of light squared.

2.0 MATTER, SPACE AND TIME

Einstein stated that nothing may travel faster than the speed of light. Due to the equivalence of energy and mass, the energy which an object has through its movement will add to its mass, making it harder to increase its speed. At 10 per cent of the speed of light an object's mass is 0.5 per cent more than normal, but at 90 per cent of light speed it would be greater than twice its normal mass. As an object approaches the speed of light its mass increases rapidly and it can never reach light speed because by then it would have infinite mass and would have needed an infinite amount of energy to get there. Only light or other such waves with no mass can travel at light speed. This indicates that the speed of light is a barrier which requires particle to wave transformation for travel at such speed.

Einstein introduced relativistic space-time where time cannot be separated from the three spatial dimensions because the rate at which time passes depends on an object's velocity relative to the speed of light and also on the strength of intense gravitational fields which have a slowing effect on time.

In 1916 Einstein published the general theory of relativity. This theory unifies special relativity and Newton's law of gravitational attraction. Gravity is considered to be a property of the geometry of space and time. The curvature of space-time is directly related to the mass-energy and linear momentum of whatever matter and radiation are present. General relativity is a metric theory of gravitation in which Einstein's field equations describe the relation between the geometry of four-dimensional space-time and the energy momentum contained in that space-time. General relativity also predicts the existence of gravitational waves which are fluctuations in the curvature of space-time, propagated as waves, travelling outward from moving objects. Energy carried by these waves is known as gravitational radiation.

Following the publication of theory of general relativity, Einstein made many unsuccessful attempts to unify gravity with electromagnetism for a unified field theory. The search for a unifying theory was halted by the discovery of weak and strong nuclear forces and by the acceptance of quantum mechanics.

The theory of everything in modern physics attempts to explain and combine all known physical phenomena. To date, theories of quantum mechanics and general relativity have not been able to be combined. In a theory of everything, gravity, electromagnetism and weak and strong nuclear forces would be unified.

The grand unified theory predicts that at extremely high energies the electromagnetic and weak and strong nuclear forces are fused into a single unified field. In the theory of everything gravity is merged into the three other gauge symmetries.

2.1.4 Quantum Mechanics

The field of quantum mechanics is of great significance in modern physics. Quantum mechanics is the set of all quantum theories which account for the dynamics of subatomic systems and has been developed from Planck's quantum principle and Heisenberg's uncertainty principle. It includes the quantisation of certain physical quantities, wave-particle duality, the uncertainty principle and quantum entanglement. Planck's quantum principle, which was expounded in 1900, stated that light or any other waves could only be emitted in discrete quanta or packets, with the energy being proportional to the frequency. In 1926 Werner Heisenberg formulated the uncertainty principle. To be able to predict the future position and velocity of a particle, its present position and velocity need to be accurately measured. However, the more accurately the position of a particle is measured, the less accurately its speed is measured. It is impossible to precisely specify both the momentum and position of a particle at the same time. Heisenberg deduced that the uncertainty in the position of a particle, times the uncertainty in its velocity, times the mass of the particle can never be smaller than a quantity known as Planck's constant. The Heisenberg uncertainty principle places limits on the accuracy with which the momentum and position of a particle can be specified simultaneously. These are fundamental limits imposed by nature and they cannot be circumvented.

Wave-particle duality is a fundamental concept in quantum mechanics and means that there is no distinction between waves and particles. Waves can exhibit particle-like behaviour and particles can exhibit wave-like behaviour. It is impossible to measure both wave and particle properties simultaneously. It was de Brolgie, who suggested in 1923 that since light waves could exhibit particle-like behaviour, particles of matter should exhibit wave-like behaviour. The wavelength of a particle is known as its de Brolgie wavelength.

In 1801 Thomas Young carried out an experiment that demonstrated the wave nature of light by showing that two overlapping light waves interfered with each other. In his experiment, known as Young's double slit experiment, two slits acted as coherent sources of light and light waves from these slits interfered constructively and destructively on the screen to produce respectively bright and dark fringes. Young used the term interference fringes to describe the bands and he deduced that these bands could only be produced if light acted like a wave. Constructive interference occurs when two waves reach the screen with the same phase displacement and produce a bright fringe. Destructive interference produces a dark fringe because waves arrive out of phase at that point.

AN INTRODUCTION TO THE NEW WORLD ORDER

At the screen light is absorbed as discrete particles known as photons. Light must have wave-particle duality which is the concept that all matter and energy have wave-like and particle-like properties.

Interference also occurs for particles. Young's experiment can be replicated by directing an electron beam onto a double slit, with the same result demonstrating that electrons also show particle-wave duality. Since the demonstrations of wave-like properties of photons and electrons, similar experiments have been completed using neutrons, protons, atoms and molecules, with these larger particles also acting as waves.

When two slits are open and a detector is added to the experiment to determine which slit a photon has passed through, two simple patterns form rather than an interference pattern. The detection of a photon involves a physical interaction between the photon and the detector. If two photons of the same frequency were emitted at the same time, they would be coherent and they would remain coherent if they went through two unobstructed slits. They would arrive at the screen at the same time but would be laterally displaced from each other and they would exhibit interference. If one or both of them encountered a detector a time delay would occur due to interaction and they would step out of phase. Because they would arrive at the screen at different times, interference could not occur. If only one photon is emitted, it would be noticed at one detector or the other, and would continue its path forward from the slit from where it was detected. The double slit experiment presents a quantum mechanics enigma in that a photon or electron can appear to be in two places at once.

Wave-particle duality indicates that everything in the Universe including light and gravity can be described in terms of particles. This infers that gravity affects light. It is known that light waves bend as they pass dense objects with gravitational redshift being a quantifiable effect. As well, photons cannot escape from black holes. A photon has zero mass but its effective inertial mass is the mass equivalent of the photon energy. The inertial mass of all bodies is equivalent to the gravitational mass.

Quantum entanglement is a property of a quantum mechanical state of a system of two or more objects in which the quantum states of the constituent objects are linked together. One object can no longer be adequately described without full mention of its counterpart. This is even though the individual objects may be spatially separated.

Entanglement is one of the properties of quantum mechanics that caused Einstein to dislike the theory.

String theory is a new developing branch of physics that combines quantum mechanics and general relativity into a quantum theory of gravity. In string theory the electrons and quarks inside an atom are one dimensional lines of vibrating energy. The implications for wave-particle duality of photons, electrons and large particles in experiments like the twin slit in terms of string theory, are significant.

2.2 Matter

Matter is defined as the substance or substances of which physical objects are composed. Within the Universe there is visible matter in the form of stars, galaxies, dust clouds, planets, asteroids and other bodies, as well as dark matter which does not reflect light because it is so dense. Black holes, which are a type of dark matter, are not visible, but their presence can be established by their anomalous gravitational influence on adjacent astral bodies.

AN INTRODUCTION TO THE NEW WORLD ORDER

A black hole is an object in which the gravitational field is so strong that its escape velocity is greater than the speed of light. Such objects form when massive stars explode. The outer layers are blasted into space and the inner layers implode to form black holes.

Mass is the amount of matter in a body. A body's resistance to acceleration or its inertia is measured by its mass. Inertia is the tendency of an object to remain at rest or in motion at a constant speed along a straight line. The mass of an object is a measure of its inertia.

Matter can be described as frozen energy. That matter and energy are interchangeable in shown in Albert Einstein's famous equation, $E = mc^2$ (energy is equal to mass times the speed of light squared). The principle of conservation of matter and energy states that matter and energy cannot be created or destroyed, but can be converted from one form to another. This means that all the matter and energy in the Universe has always existed and will continue to exist indefinitely into the future.

Einstein showed that the total energy (E) of a moving object is related to its mass and speed by the following formula:-

$$E = \frac{mc^2}{\sqrt{1 - \frac{v^2}{c^2}}}$$

When an object is at rest ($v = 0$) the total energy which is known as rest energy is:-

$$E_0 = mc^2$$

The rest energy is the energy equivalent of the mass of an object at rest. The law of conservation of mass-energy states that the sum of the mass-energy of a system of particles before interaction equals the sum of the mass-energy of the system after interaction. Therefore the total

2.0 MATTER, SPACE AND TIME

amount of mass-energy in the Universe remains the same irrespective of the type of interaction event.

The atom is a basic unit of matter. Its name comes from early Greek and means an indivisible component of matter. The classical description of an atom is a central nucleus with a positive charge surrounded by shells of electrons with a negative charge, geometrically similar to the Sun and planets in the Solar System.

The idea that all matter consists of elementary atoms differing in mass and chemical properties has been successful in explaining the many properties of matter. The periodic table was developed by the Russian chemist Dmitri Mendeleev on the basis that certain groups of elements exhibit similar chemical properties. There are eight groups with a total of 103 elements. For each element the atomic number, atomic mass and configuration of outermost electrons are given. The variety of physical and chemical properties of the elements was explained in terms of rules governing the three components, namely electrons, protons and neutrons.

Since 1945, many new particles have been discovered in experiments involving high-energy collisions between particles. The resultant particles are generally unstable with very short half lives. More than 400 temporary, unstable particles have been discovered. More recently, physicists have established that all particles, except for electrons, photons and a few related particles, comprise smaller particles called quarks. Protons and neutrons, collectively known as nucleons, consist of quarks. Electrons and quarks are currently regarded as elementary particles. There are two types of quarks known as up and down quarks. The proton is considered to comprise two up quarks and one down quark, while the neutron is composed of two down quarks and one up quark. Each down quark has an allotted charge of 1/3 of the charge of an electron and each up quark has an allotted charge of 2/3 of the charge of a proton. These charges explain the observed charges of protons and neutrons. Triplets of quarks are confined within composite particles like protons and neutrons.

Particles are subjected to four fundamental forces which, in order of decreasing strength, are strong, electromagnetic, weak and gravitational forces. Strong force binds quarks tightly together to form protons, neutrons and other hadrons. Electromagnetic force binds electrons and protons within atoms and molecules, and is about 10^{-2} times weaker than the strong force. Weak force, which causes the beta decay of nuclei and the decay of heavier quarks and leptons, has a strength of about 10^{-6} times that of a strong force. Gravitational force is a long-range force that has a strength of about 10^{-43} times that of strong force. An important aim of physicists is to discover a unified theory that will explain all four forces as different aspects of a single force.

Strong force is mediated by particles called gluons. Electromagnetic force is carried by photons (photons are the quanta of the electromagnetic field). Weak force is mediated by particles called $W^{\pm}$ and Z^0 bosons, and gravitational force is expressed by quanta of the gravitational field, termed gravitons. Field quanta like photons and gluons are collectively known as bosons, whereas entities considered to be particles such as electrons and quarks are known as fermions.

Antimatter is matter made of particles (antiparticles) with the identical mass and spin as those of ordinary matter, but with opposite charge. The positron, discovered by Carl Anderson in 1932 is the electron's antiparticle. Under proper conditions an electron and positron can annihilate each other to produce two gamma-ray photons. Antiprotons and antineutrons were produced in 1955 by Emilio Segrè and Owen Chamberlain using a particle accelerator at the University of California. Every particle has an

2.0 MATTER, SPACE AND TIME

antiparticle of equal mass and spin, of equal and opposite charge, magnetic moment and strangeness. The exceptions are the neutral photon, pion and eta, each of which is its own antiparticle. Particles and antiparticles annihilate each other. That the matter seen today comprises quarks, indicates that shortly after the big bang in our part of the Universe, symmetry T was not obeyed. This symmetry means that if the direction of motion of all particles and antiparticles is reversed, the system should revert to what it was at earlier times, i.e. the laws are the same in forward and backward directions of time. The early expansion and cooling of the big bang forces which did not obey symmetry T caused more antielectrons to turn into quarks than electrons into antiquarks. The quarks and antiquarks annihilated each other but the excess quarks remained, forming the matter that we see today.

All particles (except for photons) can be categorised as hadrons and leptons. Hadrons, which interact through the strong force, consist of mesons and baryons. Protons and neutrons are baryons. Hadrons are composed of elementary particles called quarks. Leptons are particles which occur in electromagnetic and weak interactions. This group includes the electron, muon, tau and three different neutrinos. Hadrons have size and structure whereas leptons are elementary point-like particles with no structure. Neutrinos travel close to the speed of light, have a miniscule mass, lack an electric charge and can pass through matter almost undisturbed. They form by radioactive decay or in nuclear reactions such as those that occur in the Sun. A neutrino has an antimatter partner called an antineutrino or electron neutrino, which is generated when a neutron changes into a proton or vice versa. All leptons have corresponding antiparticles. Classes of particles are included in Table 1.

AN INTRODUCTION TO THE NEW WORLD ORDER

Table 1.   Classes of Particles

Class	Subclass	Particle Name
Hadrons	Mesons	Pion Kaon Eta
	Baryons	Proton Neutron Lambda Sigma Xi Omega
Leptons		Electron Electron neutrino Muon Muon neutrino Tau Tau neutrino

The Standard Model of particle physics incorporates four basic particles in two pairs, these being the electron and neutrino and up and down quarks. The four forces are gravity, electromagnetism, and the weak and strong nuclear forces. At high energies the particle world is triplicated. The heavy counterpart of the electron is the muon with its associated muon neutrino and there are two heavier quarks called charm and strange. The heaviest electron is called the tau with its tau neutrino and the two very heavy quarks are called top and bottom. When the heavier particles are produced in accelerators they rapidly decay to first generation particles.

Bosons are particles which obey Bose-Einstein statistics in contrast to fermions which obey Fermi-Dirac statistics. Several bosons can occupy the same quantum state.

2.0 MATTER, SPACE AND TIME

Bosons are considered to be force carrier particles while fermions are generally associated with matter. Bosons may be either composite or elementary particles. There are five elementary bosons which comprise four gauge bosons (γ, g, $W^{\pm}$, $Z°$) and the Higgs boson ($H°$). The Standard Model of elementary particles is tabulated below (Table 2).

Hadrons, nuclei and atoms, which are composite particles, can be bosons or fermions depending on their composition.

Table 2. The Standard Model of Elementary Particles (after Reference 1)

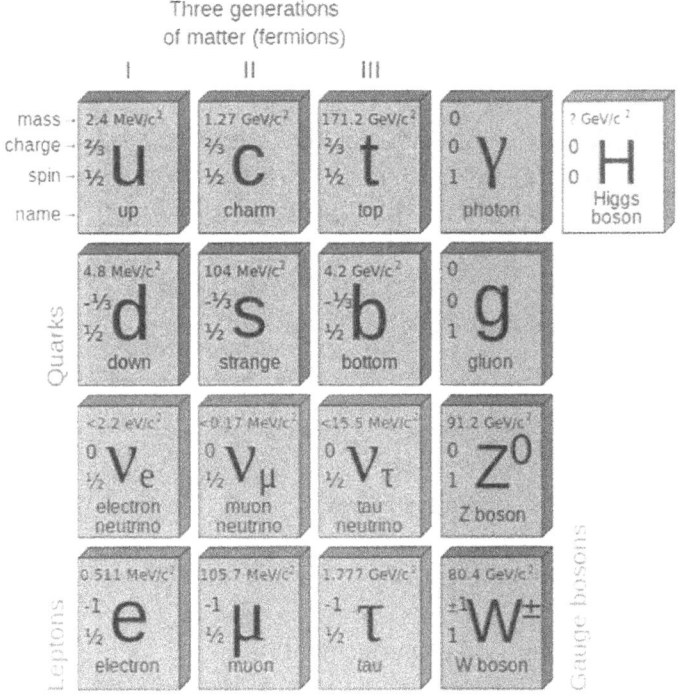

The Higgs boson is a massive elementary particle which is one quantum component of the theoretical Higgs field. It is postulated that the Higgs field has an amplitude different from zero in empty space, which gives mass to every elementary particle which should have mass. The ubiquitous, quantum Higgs field is analogous to a pool of honey that sticks to travelling, massless, fundamental particles, converting them into particles with mass which forms tangible matter, such as the components of atoms.

Supersymmetry is a symmetry that relates basic particles of one spin to another particle that differs by half a unit of spin and are known as superpartners. For every type of boson there should exist a corresponding type of fermion and conversely. Currently there is no direct evidence for supersymmetry in nature because superpartners of the particles of the Standard Model have not yet been observed.

Quantum chromodynamics is a theory of the colour force, a strong interaction, fundamental force that describes the interactions of the quarks and gluons that make up hadrons, such as the proton or neutron. It is a quantum field theory that studies colour charged fermions (quarks). There are three types of charge (red, green and blue) in quantum chromodynamics compared to two in quantum electrodynamics.

Albert Einstein stated that nothing may travel faster than the speed of light, which is the cosmic speed limit. Due to the equivalence of energy and mass, the energy which an object has through its movement will add to its mass, making it harder to increase its speed. At 10 per cent of the speed of light an object's mass is 0.5 per cent more than normal, but at 90 per cent of light speed it would be greater than twice its normal mass. As an object approaches the speed of light its mass increases rapidly and it can never reach light speed because by then it would have infinite mass and would have needed an infinite amount of energy to get there. Only light, other such waves and massless particles can travel at light speed.

2.0 MATTER, SPACE AND TIME

A new mathematical model of theoretical physics termed string theory emerged in the 1980's. String theory is in its infancy and is currently more philosophical rather than a theory of physics. Philosophical thought is often the precursor to scientific progress. String theory combines quantum mechanics and general relativity into a quantum theory of gravity. In string theory the electrons and quarks inside an atom are tiny, one dimensional lines of vibrating energy. String theory requires multiple dimensions and parallel Universes. Eventually there were five string theories in a 10 dimensional Universe and all of them appeared to be correct. In the 1990's Edward Witten proposed that the five different string theories described the same situation viewed from different perspectives. He proposed a unifying theory termed the M-theory which brought the five theories together by asserting that strings are one dimensional slices of a two dimensional membrane vibrating in 11 dimensional space.

String theory includes both open and closed strings which yield two different spectra. One of the open string modes is the photon and one of the closed string modes is the graviton. The two ends of an open string can always meet and connect to form a closed string. String theory includes objects termed branes, derived from the word membrane and include D-branes, black p-branes and Neveu-Schwarz 5-branes. Branes are extended objects that are charged sources for different form generalisations of the vector potential electromagnetic field and are related to each other by a variety of dualities. D-branes have mass and can emit and absorb closed strings. Open strings can attach to a D-brane.

The supersymmetric string theory or superstring theory is a version of the string theory that incorporates fermions and supersymmetry, unlike the bosonic string theory. In string theory space-time needs to have 10, 11 or 26 dimensions for consistency. In superstring theory there are five superstring interactions for open and closed strings. All open superstring theories also contain closed superstrings. The five superstring theories may in future be approximated to a theory in higher dimensions involving membranes. D-branes are membranes in 10 dimensional string theory. To date extending the superstring theory beyond the current 10 dimensional theory has not met with success.

A black hole is a massive object whose gravitational field is so strong that the escape velocity is greater than the speed of light. Its mass is concentrated at a point called a singularity which is surrounded by a theoretical sphere called the event horizon. Stellar black holes form when the outer layers of massive stars explode outwards and the inner layers implode. In the resulting dark matter densification which is caused by compression of protons, electrons and neutrons, forms supergravity. In black holes supergravity overwhelms all other forces and quantum uncertainty. The fabric of space-time becomes one dimensional. Supergravity compresses protons, neutrons and electrons to the densest form of matter. All external matter drawn into a black hole is similarly densified. Matter in a black hole is in relative timelessness compared to matter in the external Universe.

Black holes which are found at the centres of galaxies are called supermassive black holes. They contain hundreds of thousands to billions of solar masses and enlarge by gravitationally attracting stars and gas (black hole accretion). Intermediate size black holes contain thousands of solar masses and they probably form by collisions of black holes with subcritical masses.

Schwarzschild black holes, which have mass but neither charge nor angular momentum, are the simplest black holes. They are named after Karl Schwarzschild who discovered this solution in 1915. Other black hole

2.0 MATTER, SPACE AND TIME

solutions include charged non-rotating, rotating and stationary (having charge and angular momentum).

Black holes have mass and angular momentum. Their presence is inferred by their gravitational influence on adjacent astral bodies. The question arises as to why all matter within the Universe does not consist of dark matter. The answer lies in the effects caused by two supercritical, megamassive black holes which attract each other. Because of their mass, as they approach each other at increasing velocities due to gravitational attraction, they cannot absorb each other's dark matter. The resultant astroclysmic collision is of a big bang scale, which overcomes the entropy problem of the second law of thermodynamics, by releasing vast amounts of energy and fragmented material.

The term 'big bang' was first used by Fred Hoyle in 1949. The discovery of background cosmic microwave radiation in 1964 added credence to the big bang theory. The calculated age of the big bang is 14 billion years. A conceivable explanation for the big bang is the collision of two megamassive, supercritical black holes travelling at great velocities towards each other, in our part of the Universe.

Following the big bang, it is commonly thought that the event zone was filled with an extremely high energy density, vast temperatures and pressures and was very quickly cooling and expanding (the Planck Epoch). Rapid cosmic inflation occurred and then ceased with the impact zone consisting of a quark-gluon plasma and other elementary particles. Extreme temperatures caused the formation and annihilation of particle/antiparticle pairs. For most particles there are associated antiparticles with the same mass but opposite electric charge. Very early after the big bang, baryogenesis occurred producing asymmetry with a small excess of quarks and leptons over antiquarks and antileptons, resulting in an excess of matter over antimatter as currently occurs.

As expansion continued and temperature fell, quarks and gluons formed protons and neutrons. The temperature became too low to form new proton-antiproton and neutron-antineutron pairs. Mass annihilation occurred leaving a fraction of the original protons and neutrons and none of their antiparticles. The same process occurred for electrons and positrons. Following annihilation, the energy density of the Universe was dominated by photons and neutrinos, and protons, neutrons and electrons no longer moved relativistically.

When the temperature reached about a billion degrees Celsius, big bang nucleosynthesis occurred with neutrons and protons combining to form deuterium and helium nuclei. Cosmic microwave background radiation formed when electrons and nuclei combined to form mainly hydrogen atoms, with the radiation decoupled from matter continuing through space.

Over time gas clouds, stars, galaxies and other stellar objects formed. The three types of matter in the big bang part of the Universe are baryonic matter, hot dark matter and cold dark matter, with 72% of all matter being cold dark matter.

Matter in black holes is tightly confined by supergravity. Black hole accretion occurs by absorption of stars and galaxies, to the extent that massive, supercritical black holes occur in parts of the Universe, while other parts of the Universe become devoid of matter.

Gravitational attraction between two megamassive, supercritical black holes results in the two bodies approaching each other at ever increasing velocities, unable to absorb each other's mass. The resultant astroclysmic collision releases dark matter from supergravity, returning it to the quantum field in a huge burst of energy. Astroclysm is a universal mega-event which prevents thermodynamic heat death of the Universe. Heat death is a possible final state of the Universe where there is no thermodynamic free energy to sustain motion.

Cosmological redshift occurs when light emitted or reflected from an object is shifted towards the red end of the spectrum. The greater the redshift the further is the distance to the object. Edwin Hubble discovered an approximate relationship between redshift of nebulae and the distance to them in 1929. The relation is known as Hubble's Law. This law supports the big bang theory. The largest observed redshift is that of cosmic microwave background radiation formed about 13.7 billion years ago due to radiation decoupling from matter, following the big bang.

2.3 Space

Space can be defined as the continuous expanse extending in all directions, in which all material objects are located. Space is three dimensional. Space-time is four-dimensional space whose points are events in general relativity. An event is an occurrence at a particular point in space and time and can be specified by four co-ordinates which are three dimensional space and time. In relativity there is no real distinction between space and time co-ordinates.

Space and time can be considered absolutely or relatively. Absolute theory was propounded by Isaac Newton. In this theory space and time are real things, being containers of infinite extension or duration within which the whole succession of natural events in the World has a definite position. Things may be at rest or moving and this is not simply a matter of their relations to other objects changing. Newtonian space is an emptiness in which events occur without being influenced by space. Such space is known as flat. Newtonian space on a cosmic scale goes on forever.

In the 19th and 20th centuries mathematicians commenced to examine non-Euclidean spatial geometries in which space is considered to be curved rather than flat. In Albert Einstein's general relativity theory, space around gravitational fields deviates from Euclidean space. In 1905 Albert Einstein proposed that space and time be combined into a single entity known as space-time.

The standard space interval is the standard metre which is the distance travelled by light in a vacuum during the exact time interval of 1/299 792 458 of a second.

A wormhole is a theoretical shortcut bridge through space and time. A wormhole has two mouths and a throat which provide a traversable shortcut through folded over space-time. While there is no visual evidence for wormholes, in general relativity, space-times containing wormholes are known to be valid solutions.

2.4   Time

A basic definition of time is the movement of matter in space. From this definition examples of timelessness would include a Universe in which all matter is motionless, a Universe in which there is only space but no matter and waves travelling at infinite speed.

Newton believed in absolute time, meaning that the interval of time between two events could be accurately and unambiguously measured. Time was unrelated to space. These apparently rational views appear to fail as light speed is approached.

In the principle of Newtonian relativity, the laws of mechanics are the same in all inertial frames of reference. Newton's laws use the Galilean transformation of co-

2.0 MATTER, SPACE AND TIME

ordinates which transforms three dimensional space and time co-ordinates from one inertial system to another moving with uniform relative velocity. In the Galilean transformation of co-ordinates the fourth co-ordinate, time is assumed to be the same in both inertial systems. This situation is incorrect as the speed of light is approached.

The Galilean addition law for velocities (Galilean velocity transformation) relates instantaneous velocities of an object relative to two observers in terms of displacement and velocity only. Time is considered an absolute unit to both observers.

A fundamental premise of Newtonian mechanics is that there is a universal time scale which is the same for all observers. Einstein abandoned the assumption of simultaneity, accepting rather that a time interval measurement is dependent on the reference frame in which the measurement is made. Because there is no preferred inertial frame of reference, there is no correct answer for time events viewed from two reference frames.

Einstein's relativity theory implies that the equality of two time intervals is not intrinsic but is only relative to whichever clock is chosen to define duration. In the theory of relativity there has been a philosophical shift to the view that the simultaneity of two events does not correspond to any unique physical reality. This would be possible if two clocks spatially apart could be synchronised but this cannot be done without making assumptions about the speed of light which result in events simultaneously relative to one observer not being so relative to another observer, who is in motion with respect to the first observer. However all observers would measure the same speed of light, irrespective of their travel speed.

AN INTRODUCTION TO THE NEW WORLD ORDER

Clocks moving relative to an observer appear to be slowed down by a factor known as gamma and this is known as time dilation. The time interval measured by an observer moving with respect to the clock (C) is longer than the time interval measured by an observer at rest with respect to the clock ($\Delta t'$), because gamma (γ) is always greater than unity. Time dilation is expressed as:-

$$\Delta t = \frac{\Delta t'}{\sqrt{1-(v^2/c^2)}} = \gamma \Delta t'$$

where $\gamma = (1 - v^2/c^2)^{-0.5}$

Proper time is defined as the time interval between two events measured by an observer who sees the events occurring at the same point in space. Proper time is always measured by an observer moving along with the clock. A moving clock runs slower than a stationary clock by a factor of gamma. All physical processes in a moving frame slow down relative to a stationary clock. Time dilation is a real phenomenon which has been verified by experimentation.

The proper length of an object is the length of an object measured by someone who is at rest with respect to the object. The length of an object measured in a reference frame in which the object is moving is less than the proper length and this phenomenon is known as length contraction. Length contraction (L) is expressed as:-

$$L = L' (1 - v^2/c^2)^{0.5}$$

Where L' = proper length
and $(1 - v^2/c^2)^{0.5}$ is a factor less than one

Length contraction only takes place along the direction of motion. Proper length and proper time can only be measured in different reference frames.

2.0 MATTER, SPACE AND TIME

A consequence of time dilation is the frequency shift of light emitted by atoms in motion compared to light emitted by atoms at rest. The Doppler shift of light waves depends on the relative speed of the source and observer, and holds for relative speeds as high as c. Redshift is the Doppler displacement of spectral lines toward the red end of the visible spectrum, coming from distant stars receding from us.

Galilean transformation is not valid when velocity approaches light speed. The Lorentz transformation is a set of formulae which relates space and time co-ordinates of two inertial observers moving with a given relative speed. The time and length contraction formulae expressed above are examples of Lorentz transformation.

Newton's laws are invariant according to Galilean transformation. However the correct relativistic transformation is the Lorentz transformation. Therefore Newton's laws need to be corrected or generalised to conform to the Lorentz transformation and the principles of relativity. Generalised definitions of Newton's laws of momentum and energy reduce to classical definitions where velocity is considerably less than light speed.

Relative time anomalies which occur include the event horizons of black holes and light speed. Relativity predicts that time is apparently slower near a massive body. Hawking states that there is a singularity of infinite density and space-time curvature within a black hole. This infers that time ceases to exist when the singularity is reached.

Relativity indicates that time slows down as the speed of light is approached. Therefore at light speed there would appear to be no time.

AN INTRODUCTION TO THE NEW WORLD ORDER

The examples of black holes and travel at light speed may appear to be timeless but only with respect to specific reference frames.

Because matter and energy cannot be created nor destroyed, it is evident that all the matter in the Universe has always existed and will continue to exist indefinitely. Therefore the big bang was not the beginning of time, but an enormous explosive event, preceded by major physical changes which caused such an event.

The infinite existence of matter in the Universe and its continual movement means that the universal time interval is infinite. Big bang type events overcome the thermodynamics entropy problem.

Universal time can be represented on a number line with minus infinity representing the infinite past, zero being the present and plus infinity representing the infinite future (Figure 1). The present is a dynamic moment in time.

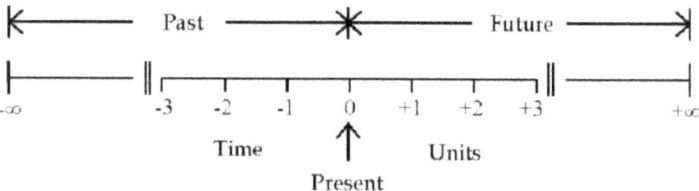

Figure 1. Universal Time Number Line.

Figure 2 shows the relationship of a traveller in relative time with respect to the universal time number line.

2.0 MATTER, SPACE AND TIME

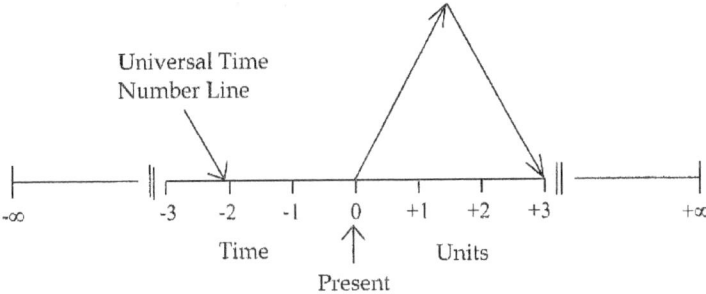

Figure 2. A Traveller in Relative Time Related to the Universal Time Number Line.

A traveller departing at a certain velocity experiences an apparent slow down in time. This relative time event has no influence on universal time and can be explained by the physical and chemical reactions in a moving object slowing down with respect to an apparently stationary object. The faster the object travels, the slower are the reactions with a resultant slower relative time frame. When the traveller returns to the universal time number line he has aged less compared to objects that have remained relatively stationary. Universal time remains unaffected.

At sub-light speeds the relationship between relative time and a reference frame time (e.g. Earth time) is expressed by the time dilation function. At and above light speed this function is not applicable. A number of different waves for a spectrum about light speed are detectable and do not appear to be travelling in a timeless environment. Waves travelling at infinite speed would experience timelessness. Quantifiable time refers to the movement of matter in space at sub-light speed.

3.0 THE UNIVERSE

3.1 General

It is commonly thought that the Universe commenced about 14 billion years ago with the big bang. An a priori argument suggests that the big bang was caused by an enormous explosive event. This means that the big bang was not the physical beginning of the Universe but rather the start of the current behavioural pattern in our part of the Universe. Because matter and energy cannot be created or destroyed but can be converted from one form to another, matter would have existed before the big bang. This could be validated if light older than the big bang is discovered and if the material that has been blown furthest from the big bang is accelerating due to attraction from matter external to the big bang zone.

There are four classical type and one relativistic possibilities for the distribution of matter in space.

1. The Universe is infinite and the amount of matter in the Universe is infinite. In this case the big bang occurred in a part of an infinitely large Universe.

2. The Universe is finite and the amount of matter within the Universe is infinite. It is impossible to have an infinite amount of matter in a finite volume of space. Questions which arise with this possibility include what constitutes the boundary of a finite Universe, can space stop at a boundary, and what is on the other side of a space boundary?

3. The Universe is infinite and the amount of matter is finite. The main problem with this possibility is that eventually the finite matter would dissipate in an

3.0 THE UNIVERSE

infinite Universe such that it would not be recognisable.

4. The Universe is finite and the amount of matter within the Universe is finite. This possibility has the same problems with universal boundaries as possibility 2. and requires a specific amount of matter for the Universe to behave as it does.

5. In relative space-time, the Universe is considered to be curved and finite with no boundaries. The amount of matter is also finite. The Universe is not thought of as expanding into pre-existing space but rather all the space that the Universe has ever had, has been in the Universe from the beginning (taken as the big bang singularity), and space is stretching as the Universe expands. As well as the problems of finite space boundaries, the requirement for the Universe to have a specific amount of finite matter to behave the way it does, space-time is dependent on the behaviour of matter within its three dimensional space. For time to occur, matter needs to move in space. A time-driven, curved universe behaving as a closed system, imposes restrictions which do not occur in an open universe model.

The extent of the Universe and the distribution of matter within the Universe is currently conjectural and more cosmic knowledge is required to improve astronomical understanding. Evidence revealed to date supports an infinite Universe containing an infinite amount of matter, with the big bang being an enormous explosive event which occurred in our part of the Universe, about 14 billion years ago.

The physical events which occurred immediately following the big bang are described in Section 2.2. The expansion of the big bang part of the Universe was predicted by Alexander Friedman in 1922. Gravitational attraction would cause a static Universe to contract. In 1929 Edwin Hubble introduced the Hubble Law, which states that the farther away a galaxy is, the larger is the redshift (Doppler effect) in its spectrum.

The big bang part of the Universe (i.e. big bang Universe) is considered to be inflationary with critical density. The inflationary theory proposes that shortly after the big bang, expansion was much more rapid than it is currently. Early expansion was exponential and current expansion appears to be linear. Critical density means that the cosmic matter density in the big bang Universe is at a value which is geometrically half way between open and closed. In an open Universe its geometry is hyperbolic which would result in expansion overcoming deceleration and the Universe would continue to expand for ever. The Universe would eventually undergo heat death which would result in black galaxies drifting outwards in to endless space. The closed Universe is a cosmological model in which the Universe is geometrically spherical and expansion would eventually stop, followed by contraction and collapse. A third possibility is that the Universe is perfectly balanced between these two alternatives so that it will continue expanding indefinitely, endlessly slowing but never quite ceasing to expand. In this possibility the Universe has a critical density.

The density of the big bang Universe is currently difficult to quantify because the total usable mass is less than one hundredth of the required amount to prevent expansion of the Universe. A large quantity of dark matter and other mass is required to prevent indefinite expansion.

The Hubble Constant which is the rate at which the big bang Universe is expanding, currently about 5 to 10 percent every thousand million years, and the deceleration parameter, which is the rate at which cosmic expansion is slowing down, are combined to give a direct index of cosmic mass density, termed omega. If omega is

3.0 THE UNIVERSE

less than one the big bang Universe will expand indefinitely, and if omega is greater than one the big bang Universe will eventually collapse. If omega is equal to one, the big bang Universe is at critical density and will expand indefinitely, at a rate approaching zero.

Background microwave radiation from big bang nucleosynthesis was first detected by Arno Penzias and Robert Wilson in 1965. This radiation is early light which has been greatly redshifted. Such evidence gives credence to the big bang theory.

The current average density of the big bang Universe is about 9.9×10^{-3} grams per cubic centimetre which is very low. The mass-energy appears to consist of 73% dark energy, 23% cold dark matter and 4% ordinary matter, with the density of atoms being about one hydrogen atom for every four cubic metres of space.

Most of the matter in the big bang Universe comprises six leptons and six quarks which interact through gauge bosons according to a particular type of gauge symmetry.

The Universe outside of the big bang Universe is unknown. If the Universe is infinite and the amount of matter in the Universe is infinite, it would be logical to assume that the average density of all of the Universe would be about the same as the average density of the big bang Universe.

Evidence that would support the presence of an infinite Universe with infinite matter includes the presence of light which is older than the big bang, matter moving out into the Universe at an accelerating rate and a consistent density of matter in the currently visible outer Universe.

3.2 Astronomy

Astronomy is the scientific study of celestial bodies. Current terrestrial and extraterrestrial telescopes study the big bang Universe. Telescopes in space have greater resolution because they are not subjected to Earth's atmospheric interference. Astronomy is of considerable significance to the human race as it provides knowledge to commence the task of successfully achieving space travel.

Cosmic microwave background radiation formed shortly after the big bang with a range of infrared to microwave wavelengths, peaking at a wavelength of about 2 millimetres. This indicates an extremely cold radiation of the blackbody spectrum. Any object that absorbs all incident radiation is called a blackbody.

Optical astronomy is the oldest form of astronomy and is known as visible light astronomy. Modern images are created using digital detectors. The Hubble space telescope has produced the most detailed images of celestial bodies.

Radio astronomy uses radio waves with wavelengths greater than about one millimetre to study celestial phenomena. Bodies which are observable at radio wavelengths include galactic nuclei, pulsars, quasars, nebulae and supernovae. In 1964 Arno Penzias and Robert Wilson discovered the radio signal left by the big bang. This is a persistent 73.5 millimetres wavelength known as cosmic background radiation.

Infrared astronomy is the detection and analysis of the Universe using infrared or heat radiation. This radiation is largely absorbed by the atmosphere. Infrared radiation allows astronomers to study dense dust clouds in the Milky Way, other galaxies and stellar nurseries.

3.0 THE UNIVERSE

The study of ultraviolet wavelengths between 10 to 320 nanometres is known as ultraviolet astronomy. Because these wavelengths are absorbed by the Earth's atmosphere, observations are required from the upper atmosphere or from space. These wavelengths are used to investigate hot, blue stars, variable stars (apparent brightness when seen from Earth varies with time), white dwarf stars, active galactic nuclei, planetary nebulae (dead stars), supernova remnants and the release of gas from comets.

X-ray astronomy is used to investigate high energy objects in the Universe such as binary stars, supernovas, elliptical galaxies, clusters of galaxies and galactic nuclei. Because X-rays are absorbed by the Earth's atmosphere, all X-ray observations are completed above the atmosphere.

Gamma-rays have the shortest wavelengths of the electromagnetic spectrum and their energies are 10 000 times greater than visible light photons. Gamma-rays are detected above the Earth's atmosphere. Most gamma-ray sources produce short term gamma-ray bursts. Continuous gamma-ray emitters include pulsars, neutron stars, supernovas, galactic nuclei and solar flares.

Cosmic rays are very high energy subatomic particles that hit gas atoms in the Earth's atmosphere, causing the atoms to fragment and producing secondary cosmic rays, which shower down on Earth. Because cosmic rays are charged, they are affected by magnetic fields and so their sources have not been identified.

3.3 Interstellar Medium

Interstellar medium is the gas and dust between the stars of a galaxy and consists of a sparse mixture of ions, atoms, molecules, dust particles, cosmic rays and magnetic fields. The matter consists of about 99% gas and 1% dust by mass with densities ranging from several thousands to several hundred millions of particles per cubic metre.

Interstellar medium comprises about 10% of the Milky Way's mass and is concentrated in the spirals and the central disc of this galaxy. The gas is about 89% hydrogen, 9% helium and 2% of heavier elements. The interstellar medium is turbulent and full of structure.

3.4 Molecular Cloud

A stellar molecular cloud or nursery is a cool, dense region of interstellar matter with a density about a million times greater than interstellar medium. Molecular clouds comprise hydrogen molecules and cosmic dust.

The largest molecular clouds, termed giant molecular clouds, are up to 10 million times the Sun's mass. Because giant molecular clouds are so large and cover significant parts of constellations, they are often named after the constellations in which they occur.

Small molecular clouds are isolated, gravitationally bound small clouds with masses less than several hundred times the mass of the Sun. High-latitude, diffuse molecular clouds are diffuse filamentary clouds which occur at high galactic latitudes.

New stars form within molecular clouds due to their low temperatures, high densities and gravitational forces, causing collapse of the matter in the clouds.

3.5 Nebula

A nebula is a high temperature, interstellar cloud of dust, hydrogen, helium and other gases. Many nebulae form from the gravitational collapse of gas in the interstellar medium.

A reflection nebula is a dust cloud which reflects light from a nearby star or stars. A dark nebula is a dust cloud which blocks light. Planetary nebulae are gas shells emitted by some stars at the ends of their lives. Some nebulae form as a result of their supernova explosions which occur at the death of massive, short-lived stars.

3.6 Protostar

Protostars are the earliest stage of stellar evolution and form by the contraction of gas in giant molecular clouds. The protostellar phase which lasts for about 100 000 years starts with a core of increased density, ending as a T Tauri star. T Tauri stars have central temperatures which are too low for hydrogen fusion and are powered by gravitational energy. These stars contract to form main sequence stars after a period of about 100 million years. A super solar wind, known as the T Tauri wind marks the change from an accreting star to one that radiates energy.

3.7 Stars

3.7.1 Definition

A star is a luminous sphere of plasma which is contained by gravity. Stars radiate light through thermonuclear fusion with nearly all elements heavier than hydrogen and helium being formed by fusion in stars.

A star commences as a collapsing cloud of mainly hydrogen and when the core is dense enough, nuclear fusion causes hydrogen to be converted into helium. Internal pressure prevents further gravitational collapse. When the core hydrogen is exhausted, stars which have at least 0.4 times the mass of the Sun expand to become red giants. These stars then become degenerate forms.

3.7.2 Star Classification

Stars are classified according to their colour, temperature and size. Most stars are divided into one of seven spectral types (O, B, A, F, G, K and M) with each type being divided into subclasses ranging from 0 for the hottest to 9 for the coolest stars. Stars of a specified spectral type are further subdivided into luminosity classes which are size related. These classes are Ia (bright supergiants), Ib (supergiants), II (bright giants), III (giants), IV (subgiants) and V (main sequence dwarfs).

The Hertzsprung–Russell Diagram (named after its co-developers) relates the temperatures and luminosities of stars (Figure 3). The stars in the upper left of the diagram are young, hot, giant blue stars and those in the upper right are cooler red giants and supergiants near the ends of their life cycles. Stars located in the lower left of the chart are white dwarfs which are very faint, hot, small stars at the ends of their lives. Stars spend most of their lives on the main sequence band. Main sequence or dwarf stars form about 90% of all stars.

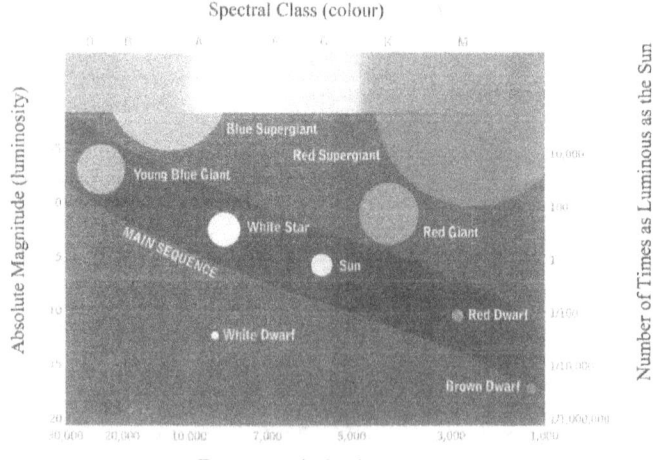

Figure 3. Hertzsprung-Russell Diagram (after Reference 2).

3.0 THE UNIVERSE

3.7.3 Main Sequence Star

All main sequence stars are in hydrostatic equilibrium. In stars below 1.5 times the mass of the Sun, hydrogen atoms fuse in a series of stages known as the proton-proton chain to form helium. Above this mass, the fusion process uses atoms of carbon, nitrogen and oxygen as intermediaries in the carbon-nitrogen-oxygen cycle to produce helium from hydrogen.

The more massive the star, the shorter is its lifespan because massive stars have greater pressure on their cores, resulting in hydrogen being more rapidly burnt. Most stars are between 1 billion and 10 billion years old with the oldest star yet discovered being 13.2 billion years old.

Main sequence stars spend about 90% of their lives fusing hydrogen to helium. As the proportion of helium increases in the cores, stars will slowly increase in temperature and luminosity. The Sun is estimated to have increased in luminosity by about 40% since it became a main sequence star 4.6 billion years ago.

The duration that a star spends in the main sequence band depends on its mass and luminosity. The Sun, which is a yellow dwarf, has been a main sequence star for about 4.6 billion years and it will become a red giant in about 5 billion years.

Stars of at least 0.4 solar masses expand greatly and cool to form red giants when their core hydrogen supply is exhausted. In red giants of up to 2.25 solar masses, hydrogen fusion continues in shell layers surrounding the cores which are eventually compressed enough to commence hydrogen fusion. This results in these stars shrinking in size and increasing in surface temperature.

In larger stars, helium fusion directly replaces hydrogen fusion in the cores.

Following helium fusion in the cores, helium fusion continues in the shells around hot cores of carbon and oxygen. Stars with more than nine solar masses expand to form red supergiants. Once all helium has been used in the cores, heavier elements continue to fuse.

The cores contract until the temperature and pressure are sufficient to successfully fuse carbon, neon, oxygen and silicon forming layered shells within the stars. In the final stages iron is produced without releasing energy and in very massive stars, accumulation of large cores of inert iron occurs. Heavier elements in these stars can move towards the surfaces forming Wolf-Rayet stars which have dense stellar winds.

A red dwarf star is a small, cool, main sequence star with a mass of between 0.6 and 0.1 solar mass. Red dwarfs comprise about 70% of all stars. Red dwarf stars have relatively low core temperatures with energy being produced at a slow rate through nuclear fusion of hydrogen into helium. In red dwarfs energy is conveyed from the core to surface by convection. These stars emit little light and because they consume nuclear fuel very slowly, they can live for about 100 billion years.

Brown dwarfs are sub-stellar bodies with masses which are too low to sustain nuclear fusion.

3.7.4 Post-Main Sequence Events

In an average size star, in the post-main sequence phase, a planetary nebula forms when the star can no longer support itself by nuclear fusion, and gravity forces the inner part to condense and heat up. The hot centre causes the outer layers to be shed in a stellar wind resulting in a white dwarf star with less than 1.4 solar masses. The degenerate matter in a white dwarf is matter compressed so tightly that electrons are stripped from their nuclei and free electrons and nuclei exist in a closely packed mass. White dwarfs eventually fade into black dwarfs over considerable time.

A nova is a cataclysmic nuclear explosion caused by the accretion of hydrogen on the surface of a white dwarf star, which ignites and begins uncontrolled nuclear fusion. The white dwarf accretes gas from the outer atmosphere of a close companion star. The vast amount of energy released in this process blows the remaining gases away from the surface of the white dwarf, producing an extremely bright outburst of light. A white dwarf can produce recurrent novae as it continues to accrete surface hydrogen from its companion star.

In large stars fusion continues until the iron core has enlarged such that it can no longer support its own mass. Sudden core collapse causes electrons to be driven into its protons, forming neutrons and neutrinos in a burst of inverse beta decay. The shockwave caused by this sudden collapse causes the entire star to explode in a supernova, which is an extremely luminous stellar explosion, and can produce a radiation outburst which can briefly outshine an entire galaxy. After most of the matter in the star has been blown away, the remaining mass exists as a neutron star or a pulsar. In the largest stars with stellar remnants greater than about four solar masses, black holes are formed.

3.7.5 *Neutron Star*

A neutron star is the smallest and densest star with a mass of about 1.4 to 3 times the Sun's mass. A neutron star begins as a main sequence star with about ten solar masses. As a neutron star approaches its end, gravitational collapse causes electrons and protons to collapse into neutrons. Neutron stars are extremely densely packed and are not dense enough to form black holes.

3.7.6 Pulsar and Magnetar

A pulsar is a rapidly spinning neutron star which transmits considerable energy. The spin is so fast that its strong magnetic field generates a surrounding electric field and a beam of radiation is generated at the north and south magnetic poles. A young pulsar spins very rapidly but as the magnetic field weakens with age, the rotation slows, extinguishing the radiation beam.

A magnetar is a young neutron star with a magnetic field from 100 to 1000 times stronger than a standard neutron star. A magnetar produces recurring bursts of low-energy gamma-rays.

3.7.7 Binary Stars

Binary and multi-star systems consist of two or more stars that are gravitationally bound and they move around each other in stable orbits.

3.7.8 Constellation

A constellation is a group of stars which appear to form a pattern in the sky. Astronomers currently focus on constellations as grid-like segments of the celestial sphere rather than as patterns. A star pattern such as the Big Dipper, which is not officially classified as a constellation, is known as an asterism.

3.7.9 Galaxy

A galaxy is a vast aggregation of stars, gas, dust and dark matter bound together by gravitational attraction. The three main galaxy shapes are elliptical, spiral and irregular.

3.0 THE UNIVERSE

Elliptical galaxies are round or elliptical and have virtually no internal structure. They lack nebulae and hot bright stars but older red giant stars are common. Stars in elliptical galaxies all follow individual orbits around the galaxies' cores, which are probably supermassive black holes.

Spiral galaxies are either basic spirals or barred spirals. The latter have elongated nuclear bulges. Spiral galaxies contain old and young, hot and cool stars as well as gas and dust that promote star formation along the spirals. The inner curves of the spirals' arms consist of bright nebulae and dark dust lanes.

Irregular galaxies contain various stars and large clouds of gas and dust. These galaxies have no distinct shape, symmetry or nucleus and have a number of ionised hydrogen zones.

Interacting galaxies are galaxies which are gravitationally merging, resulting in intense star formation and distortion of galaxy structures including spirals. In some cases larger galaxies entirely consume smaller galaxies.

Radio galaxies are the largest class of galaxies and are generally elliptical galaxies. They emit intense radiation at radio wavelengths. Very bright radio galaxies have similar characteristics to quasars.

Seyfert galaxies have starlike nuclei that emit enormous amounts of highly ionised energy. These galaxies are termed Type 1 Seyfert galaxies and they behave like quasars. Type 2 Seyfert galaxies are bright in the infrared wavelengths, having much narrower emission lines. Most Seyfert galaxies are spiral galaxies. About 25% of Seyfert galaxies are irregularly shaped, indicating that they have resulted from intergalactic collisions or interactions. The luminosity of Seyfert galaxies can change by up to 50% within several days or months. This is probably because these galaxies have supermassive black holes at their cores which result in sudden increases in brightness as matter surges into these black holes.

3.7.10 Quasar

A quasi-stellar radio source or quasar is a very active and distant galactic nucleus. Quasars, which show very high redshift, are some of the most distant objects in the Universe. They emit enormous amounts of energy, commonly 10 000 times the energy of an average spiral galaxy. A quasar is a core of a massive galaxy where large quantities of matter are flowing into a central supermassive black hole from which light cannot escape. Released energy is generated outside the event horizon by gravitational stresses and enormous friction on incoming material, which forms an accretion disc around the black hole. Quasars were much more common shortly after the big bang.

Quasars can vary in brightness over short time frames down to hours, due to changes in the amount of released gravitational energy from matter being absorbed. When matter stops flowing in, the quasar becomes a normal galaxy. Quasars can regenerate when infused with new matter. It is possible for a quasar to form when the Milky Way Galaxy collides with the Andromeda Galaxy in about 3 to 5 billion years.

3.7.11 Globular Cluster

A globular cluster is a spherical group of stars that orbits a galactic core. Such clusters which are tightly bound by gravity, have high stellar densities towards their centres.

3.7.12 Galaxy Groups, Clusters and Superclusters

Galaxy groups and clusters are the largest gravitationally bound objects in the cosmos. Most matter in the

Universe occurs as filaments or clumpy areas which can contain thousands of galaxies. Galaxy clusters have formed in the last 10 billion years. Super clusters are larger non-gravitationally bound groups of galaxy clusters.

Galaxy clusters have large quantities of dark matter. Groups of galaxies can move at velocities to about 800 km/sec.

3.8 The Solar System

The Solar System comprises the central Sun surrounded by celestial objects which are gravitationally bound in orbits. This system formed from the collapse of a giant molecular cloud about 4.6 billion years ago. The cloud comprised a vast, cold, rotating mass of interstellar gas and dust known as the solar nebula. Gravity caused most of the gas and dust to centralise forming a dense mass known as the protosun which continued to condense and started to vaporise ice in the central region. The rest of the spinning nebula began to flatten into a disc with lighter gases such as hydrogen and helium being pushed to the outer regions.

Fragments of rock and dust within the nebula began to accrete to form planetesimals and then protoplanets. In the inner regions terrestrial planets formed whereas in the outer, cooler regions the planetesimals formed the stony cores of the giant planets which accumulated thick, gaseous atmospheres.

The protosun commenced as a T Tauri star which is a very young, irregular, pre-main sequence, variable star. In these stars the central temperatures are too low for hydrogen fusion and they power by release of gravitational energy as they contract. After about 50 to 100 million years the pressure and density in the protosun became sufficient for the start of nuclear fusion, and hydrostatic equilibrium was achieved, whereby thermal energy counteracted the force of gravitational attraction. The Sun became a yellow dwarf, main sequence star.

Mercury, Venus, Earth and Mars which are the four smaller inner planets, are composed of rock and metal, and are called the terrestrial planets. The four outer planets are gas giants. Jupiter and Saturn are the two largest planets and they are mainly composed of hydrogen and helium. Uranus and Neptune comprise largely ice, being frozen water, ammonia and methane, and are known as ice giants.

Two regions of smaller objects occur in the Solar System. The Asteroid Belt which occurs between Mars and Jupiter comprises mainly rock and metal fragments. Beyond Neptune's orbit is the Trans-Neptunian Region or the outer Solar System which is composed mainly of ice and rock. Within these two regions, five bodies occur which have been rounded by their own gravity, and they are called dwarf planets. Ceres occurs in the Asteroid Belt and Pluto, Haumea, Makemake and Eris are found in the Trans-Neptunian Region. Six of the planets and three of the dwarf planets are orbited by natural satellites or moons. Planetary rings of dust and ice encircle the outer planets.

The solar wind which is a plasma flow from the Sun creates a bubble in the interstellar medium which is called the Heliosphere. There are six planets in the Solar System which have a magnetosphere, being the area around a planet that is occupied by its magnetic field. The planets are Mercury, Earth, Jupiter, Saturn, Uranus and Neptune. The Earth's magnetic field prevents its atmosphere from being stripped away by the solar wind. Because Venus and Mars do not have magnetic fields, the solar wind causes their atmospheres to gradually dissipate into space. The solar wind is a stream of atomic particles, mainly protons and electrons, emanating from the Sun's corona, at a rate of about 1 million tonnes of matter per second. Wind velocity ranges from 450 to 900 km per second, with the fastest speeds being at coronal escape holes. Solar winds cause auroras and make the tails of comets point away from the Sun. These winds reach the Heliopause which is the start of the interstellar medium.

3.8.1 The Sun

The Sun is a yellow dwarf star which is at the centre of the Solar System. It has a diameter of 1 392 000 km. About 75% of the Sun's mass is hydrogen with the most of the rest being helium. The Sun generates its energy by nuclear fusion of hydrogen to form helium. The mean distance between the Sun and the Earth is 149 597 871 km (1 Astronomical Unit or 1 AU).

The photosphere is the visible surface layer of the Sun. Above the photosphere is the solar atmosphere which consists of five zones being the temperature minimum, the chromosphere (a dim layer of gas about 2000 km thick), the transition region, the corona (an extensive, wispy, hot upper layer) and the Heliosphere (the Sun's area of magnetic influence extending about 50 AU to 100 AU from the Sun, and where the solar wind pushes back the interstellar medium). The photosphere is tens to hundreds of kilometres thick and is slightly less opaque than air on Earth.

The convective zone is the Sun's outer layer which extends from the photosphere down to a depth of 200 000 km. Because the solar plasma in this zone is not sufficiently dense or hot enough to radiate thermal energy, hot material is transferred to the surface by thermal convection.

The radiative zone which occurs from about 0.25 to 0.7 solar radii, is where solar material is sufficiently hot and dense to be able to radiate energy by transfer of photons from hydrogen and helium ions, to be reabsorbed by other nearby ions.

The solar dynamo is the physical process that produces the Sun's magnetic field which reverses itself about every 11 years, causing sunspot activity. Magnetic field generation occurs at the tachocline, which is the transition region between the radiative and convection zones. This area has a very large shear profile which forms strong magnetic fields.

The Sun's core, which extends from the centre to about 0.3 of the solar radius, is very dense and hot. Considerable amount of thermal energy is produced in this zone through hydrogen fusion.

Sunspots are regions in the Sun's photosphere where a very strong magnetic field slows gas rising to the surface. The centre of a sunspot is depressed and cooler than the surrounding gas. Sunspots often occur in clusters and increase during the 11 years magnetic cycle.

Solar flares are sudden eruptions on the Sun's surface and are prevalent during the peak of sunspot activity. They eject energy, billions of tonnes of charged particles and radiation in the radio waves to X-rays range. Solar flares cause auroras and geomagnetic storms on Earth.

The Sun is about halfway through its main sequence phase. Because the Sun does not have enough mass to explode as a supernova, in about 5 billion years it will become a red giant. The outer layers will expand as hydrogen fuel in the core is depleted. The core will contract and heat up resulting in helium fusion when the temperature reaches about 100 million degrees Celsius. The Earth will be engulfed in the red giant Sun resulting in evaporation of all water and destruction of the atmosphere.

After the red giant phase the Sun's outer layers will disperse and a planetary nebula will form. The very hot core will cool and fade as a white dwarf over many millions of years.

3.8.2 Mercury

Mercury, the closest planet to the Sun, is also the smallest planet and has no moons. Its solar orbit is highly elongate. It is very dense, probably with a large iron-rich core. Surrounding the core is a mantle to 700 km thick, consisting of silicates. The outer, lighter layers were probably removed by an early giant impact. Surface topography comprises flat plains, lobed ridges and impact craters. Mercury is too hot to maintain an atmosphere and the surface is blasted by the solar wind. The average temperature is 167°C.

Mercury has a strong magnetic field capable of deflecting the solar wind around the planet, forming a magnetosphere.

3.8.3 Venus

Venus is the second planet from the Sun and is similar in size to Earth, but smaller. The brightness of Venus (second to the Moon) is due to a blanket of reflective clouds indicating a thick atmosphere. Venus has no moons and it is the hottest planet with a mean surface temperature of 464°C, due to abundant atmospheric, greenhouse gases, comprising 96% carbon dioxide. The surface is smooth and flat, probably due to extensive lava flows. There are few craters and thousands of volcanoes occur on the plains.

Venus has a thick siliceous mantle which surrounds an iron core. It has no significant magnetic field to maintain an atmosphere in prevailing solar winds. The dense atmosphere probably occurs due to constant replenishment by gases from volcanic eruptions. The small induced magnetosphere provides negligible protection to the atmosphere from cosmic radiation.

3.8.4 Earth

The third planet from the Sun is Earth which is the largest and densest of the inner planets. About 71% of the surface is covered with water and the nitrogen and oxygen atmosphere supports life. Earth's axis is tilted at 23° to the plane of the ecliptic resulting in distinct seasons.

Earth has a solid iron inner core and a liquid iron outer core, which is surrounded by the mantle. The crust is thin and brittle, comprising moving plates which form mid-ocean ridges, rift valleys, subduction zones and mountain chains.

Earth's oceans and atmosphere are essential for life. Evolution of life has been accompanied by explosive evolution phases and mass extinctions.

Earth's magnetic field is generated by the heat of the inner core which causes convection currents in the liquid outer core. The magnetic poles constantly move. The magnetic polarity reverses at periods ranging from tens of thousands to hundreds of thousands of years. The magnetosphere protects the Earth from the particles of the solar wind which are electrically charged, and are trapped by the magnetic lines of force in two large rings called the Van Allen Belt. The inner belt which is about one Earth radius above the Equator contains high energy protons and electrons. The outer belt located about three Earth radii above the Equator consists mainly of lower energy electrons. The Aurora Borealis or northern hemisphere lights and Aurora Australis or southern hemisphere lights are caused by showers of charged particles that escape from the Van Allen Belt and descend into the ionosphere causing gas fluorescence. Brilliant auroras occur during geomagnetic storms when the

3.0 THE UNIVERSE

Sun's activity is greatly increased by solar flares or coronal mass ejections.

The biosphere comprises the parts of Earth that contain life, which evolved in the ocean over 3 billion years ago. The atmosphere is a layer of gases which consists mainly of nitrogen and oxygen. It is about 600 km thick and can be divided into five zones. The troposphere is the lowest level of the atmosphere and extends upwards for about 10 km. The next layer is the stratosphere which extends to about 50 km. The stratosphere contains jet streams and absorbs most the Sun's ultraviolet radiation. The mesosphere is directly above the stratosphere and extends from about 50 to 85 km above Earth, with the temperature decreasing with height. The thermosphere extends to up to about 600 km and X-rays and other short wave radiation from the Sun energise sparsely distributed molecules. The exosphere is the upper limit of the Earth's atmosphere with thin concentrations of hydrogen and helium.

The ionosphere is the region of Earth's upper atmosphere from about 60 km to 600 km altitude and begins in the upper part of the mesosphere, extending through the thermosphere and into the exosphere. It is the zone where the Sun's radiation ionises molecules so that they reflect high frequency radio waves back to Earth. The ionosphere consists of the D, E, Es, F_1 and F_2 layers which have varying roles related to the presence of day and night.

3.8.5 *Earth's Moon*

The Moon which is at an average distance of 384 400 km from Earth, has a diameter of 3476 km and it is one quarter the size of Earth. It has a synchronous rotation, meaning that it rotates on its axis at the same rate as it orbits Earth. Its average temperature is -20°C.

Several hypotheses have been suggested for the Moon's formation at about 4.5 billion years ago, some 50 million years after the origin of the Solar System. One option was the fission of the Moon from the Earth's crust through centrifugal force which was not feasible because of the excessive initial spin required for Earth. The gravitational capture of a pre-existing Moon would not have been physically possible. Contemporaneous formation of the Moon and Earth in the primordial accretion disc does not explain the geochemical differences between the two bodies. The most plausible explanation is that the Earth and Moon formed through a giant impact where a planet termed Theia, which was about the size of Mars collided with Earth and blasted material from Earth into orbit where it accreted to form the Moon. The geochemical compositions of the Earth and Moon are compatible with this scenario.

The Moon has internal structure which comprises a solid iron inner core with a radius of about 240 km, a fluid iron outer core with a radius of 300 km, a partially molten boundary layer with a radius of about 500 km which occurs around the core, a solid mantle and a crust which is about 50 km thick.

The Moon has an airless, dry rocky surface whose main features are craters and maria, which are ancient lava flows. During its early years the Moon went through a period of intense bombardment resulting in the many craters which are currently visible. Most of the Moon's mare basalts formed lava flows from 3 to 3.5 billion years ago. Younger eruptions have been documented.

Although liquid water cannot occur on the lunar surface, surface water ice has been detected. The Moon's surface gravitational field is about 0.17 that of the Earth's field. The Moon has an external magnetic field but does not have a global dipolar magnetic field like Earth. The Moon's atmosphere is rarefied enough to be classified as a vacuum. Common atmospheric molecules are absent, probably the result of solar activity.

3.8.6 Mars

Mars, which has a distinctive reddish colour, is the fourth planet. Its colour results from the presence of iron oxide in the soil. It is the second smallest planet in the Solar System, has an equatorial diameter of 6794 km, a mass about one tenth of Earth and a gravitational field of 0.38 of that of Earth. Mars has no global magnetic field, probably because of the extent to which the iron core has cooled.

The Martian atmosphere which consists mainly of carbon dioxide, nitrogen and argon has a surface pressure of 6.1 millibars which is about 0.6% the pressure of Earth's atmosphere. The average temperature on Mars is -65°C.

The Martian surface can be divided into two different halves. The southern hemisphere is old, rough and heavily cratered while the northern hemisphere has a smoother, younger terrain with many volcanoes. From 4.5 billion years ago to 3.5 billion years ago the surface of Mars was scarred by many large impact craters. Between 3.5 billion years ago and 3 billion years ago extensive lava plains were extruded. From that period some impact craters formed and geological activity still continues. Research indicates that 4 billion years ago the northern hemisphere of Mars was hit by an object about one tenth to two thirds the size of Earth's moon. Mars lost its magnetosphere about 4 billion years ago, allowing solar winds to strip away the atmosphere.

Geological features such as deltas and alluvial fans indicate that warmer, wetter conditions prevailed in earlier times. Evidence occurs for an ancient Martian ocean.

There are two permanent, polar ice caps on Mars. They comprise water ice and frozen carbon dioxide. The ice caps are vast containing millions of cubic kilometres of ice.

Mars has two small moons named Deimos and Phobos. They are considered to be captured asteroids.

3.8.7 Asteroid Belt

The Asteroid Belt which occupies the orbit between Mars and Jupiter, comprises numerous irregular shaped bodies termed asteroids which are either stony or metallic. The asteroids range in size from hundreds of kilometres in diameter to microscopic. The four largest asteroids are Ceres, Vesta, Pallas and Hygiea. Ceres has a 950 km diameter and Vesta, Pallas and Hygiea have diameters of greater than 400 km. These four asteroids have about half of the mass of the entire belt. Although the Asteroid Belt contains millions of fragments over one kilometre diameter, the total mass is less than one thousandth of the Earth's mass. The main belt has a sparse distribution of asteroids and spacecraft routinely pass through without incident.

The Asteroid Belt is considered to be one of the remnants of the precursors to planets, which subsequently formed protoplanets. These fragments failed to coalesce due to adjacent Jupiter's gravitational interference. The current Asteroid Belt contains only a small fraction of the original mass which was probably equivalent to Earth. Gravitational effects caused much of the mass to be ejected from the belt within a million years of formation.

3.8.8 Jupiter

Jupiter is the fifth and largest planet in the Solar System. It is equivalent to 318 Earth masses and is 2.5 times the mass of all the other planets combined. It is mainly composed of hydrogen and helium. It is considered to comprise a dense core with various elements surrounded by a layer of liquid metallic hydrogen with some helium and an outer layer of gaseous hydrogen. The average temperature is -110°C, with temperature and pressure increasing steadily towards the core.

The Great Red Spot, discovered in 1831 is about 26 000 km by 14 000 km in size. It is considered to be an enormous storm comprising a high pressure vortex wedged in the upper atmosphere between two jet streams. Heat from deep within the planet supplies a steady stream of energy for this permanent storm.

Jupiter's magnetic field is 14 times as strong as that of the Earth. The four largest moons are within the magnetosphere which protects them from the solar wind.

Jupiter is continuously covered with clouds of ammonia crystals and ammonium hydrosulphide which extend upwards for about 50 km. Intense storms comprising cyclones, anticyclones and thunderstorms disrupt Jupiter's clouds. Strong, steady jet streams occur in the upper atmosphere.

There are 67 satellites orbiting Jupiter. The four largest are Ganymede, Io, Europa and Callisto which are known as Galilean moons. There are eight inner, regular moons which include the Galilean moons and the remainder are irregular moons. Ganymede is the largest moon in the Solar System with a radius of 2631 km, which is larger than Mercury. The upper layers may contain significant ice and this moon has a global magnetosphere. Europa is the second closest of the Galilean moons and has a smooth shiny surface which could be water ice. Io is the closest moon and is of similar size to the Earth's moon. It has intense volcanic activity caused by the gravitational effects of adjacent bodies. Of the four Galilean satellites Callisto is the outermost one. It is the third largest moon in the Solar System and is tectonically quiescent. Numerous craters are present on the surface.

Because of its size Jupiter receives the most frequent impacts in the Solar System, which helps to shield the inner planets.

3.8.9 Saturn

Saturn which is the sixth planet, is the second largest planet and is characterised by an extensive ring system. It consists mainly of hydrogen and helium, with a mass density less than water. It is considered to have a core consisting of iron-nickel and rock, surrounded by a thick layer of metallic hydrogen, an intermediate layer of liquid hydrogen and helium and an outer gaseous layer.

The outer atmosphere of Saturn consists mainly of hydrogen and helium. Clouds comprise ammonia ice and water ice. The winds on Saturn which peak at about 500 m/sec are the second fastest of the Solar System's planets, after Neptune. The average temperature is -140°C.

Saturn has a magnetosphere probably caused by currents in the liquid metallic-hydrogen layer. The magnetosphere deflects solar winds and produces auroras.

Saturn has seven rings, which are about 20 m thick and are composed mainly of water ice. They are believed to be the remains of pulverised moons, comets and asteroids.

Saturn has 62 satellites of which Titan is the largest and it is the second largest satellite in the Solar System. It has a radius of 2575 km which is larger than Mercury and Pluto. It has a dense atmosphere consisting of mainly nitrogen with minor amounts of methane, ethane and acetylene. The average temperature is -178°C. The geochemistry is similar to that of early Earth. Rhea is

Saturn's second largest moon. Four of Saturn's small moons are significant in shaping the planet's rings. Pan, Atlas, Pandora and Prometheus are called shepherd moons because their gravitational fields help to separate and confine the particles in Saturn's rings.

3.8.10 Uranus

Uranus is 14 times as massive as Earth and it is the seventh planet from the Sun. It is the lightest of the outer planets and it is half way between the giant planets of Jupiter and Saturn. It is unique in that it orbits the Sun on its side with its axial tilt being 99.77° to the ecliptic. It is postulated that the planet was displaced by a huge collision early in its history.

Uranus's atmosphere consists mainly of hydrogen and helium. It has an average temperature of -215°C. Its magnetosphere is similar in strength to that of Earth. The planet's shell which is about 8000 km thick consists of liquid hydrogen which encases an ocean of compressed water, methane and ammonia about 10 000 km in depth. The core comprises molten rock.

Uranus has 13 thin rings composed of very dark particles which vary in size, from micrometres to boulder size. They may be coated with sooty carbon dust. The rings may be part of satellites that were shattered by high speed impacts.

There are 27 moons circling Uranus. The five largest are Miranda, Ariel, Umbriel, Titania and Oberon.

3.8.11 Neptune

Neptune is a similar gas giant to Uranus with a mass 17 times greater than Earth. It is the eighth and farthest Solar System planet from the Sun. Its atmosphere consists mainly of hydrogen and helium. The weather is characterised by extreme storms with wind speeds reaching 600 m/sec. The average temperature is -214°C but the stratosphere of Neptune is warmer than that of Uranus due to elevated concentrations of hydrocarbons. The magnetosphere is similar to that of Uranus.

Neptune's internal structure resembles the Uranus zonation. The mantle comprises mainly water, ammonia and methane. The core consists of molten iron, nickel and silicates.

Neptune has a planetary ring system comprising three main rings. The rings are probably composed of ice particles coated with silicates or carbon.

There are 14 known moons orbiting Neptune. The only large satellite is Triton which is about 75% of the size of Earth's Moon. It is geologically active with geysers of liquid nitrogen and ice volcanoes activated by the Sun's heat. Triton has a retrograde orbit in which it circles the planet in the opposite direction to the planet's rotation. The orbit is decaying and in about 250 million years Triton will disintegrate into another ring.

3.8.12 Trans-Neptunian Region

The area beyond Neptune is known as the Trans-Neptunian Region or the outer Solar System. It comprises many small objects made of rock and ice. The Region has several zones.

3.8.12.1 Kuiper Belt

The Kuiper Belt is the first formation extending beyond the planets from the orbit of Neptune at 30 AU to 50 AU from the Sun. There are over 100 000 Kuiper Belt objects with a diameter greater than 50 km. Most objects consist of frozen volatiles such as methane, ammonia and water. The largest objects are the dwarf like planets such as Quaoar, Varuna and Orcus. The three classified dwarf planets in this zone are Pluto, Haumea and Makemake.

The Kuiper Belt consists of planetesimals which are fragments from the protoplanetary disc around the Sun that failed to coalesce into planets.

Pluto is the largest object in the Kuiper Belt and it is a dwarf planet with a diameter of 2390 km. It has a thin atmosphere comprising nitrogen, methane and carbon monoxide. The average temperature is -225°C.

Pluto has five moons named Charon, Nix, Hydra, Kerberos and Styx. Charon is the largest moon with a diameter of 1190 km. It is covered in water ice and is in a mutually synchronous orbit with Pluto where they orbit a barycentre of gravity above their surfaces. They appear to orbit each other.

The dwarf planets Pluto, Haumea and Makemake are 39 AU, 43 AU and 46 AU from the Sun.

3.8.12.2 Scattered Disc

The Scattered Disc has sparsely distributed objects extending to 150 AU from the Sun. The objects have highly elliptical orbits which appear to have been gravitationally influenced by Neptune's early movement.

Eris which is a dwarf planet is the largest object in the Scattered Disc. It is about the same size as Pluto and has a highly elliptical orbit with an average distance of 68 AU from the Sun.

3.8.12.3 Heliosphere

The Heliosphere is that region of space which is dominated by the Sun. The solar wind travels in the Heliosphere at about 400 km/sec until it collides with the interstellar medium consisting of hydrogen and helium plasma which is dispersed throughout the Milky Way Galaxy. The termination shock which occurs at a distance of at least 80 AU from the Sun, suddenly slows down motion. A great oval structure termed the Heliosheath is formed beyond the terminal shock. This transitional zone is where the solar wind is slowed down, compressed and made turbulent by its interaction with the interstellar medium. The outermost boundary of the Heliosheath is called the Heliopause. This is where the solar wind and interstellar medium pressures are balanced. As the Heliopause is reached the magnetic field intensity doubles and there is a significant increase in cosmic rays. The Heliotail is the tail of the Heliosphere which is similar to a comet's tail.

Outside of the Heliosphere the plasma density increases by about 40 times. Beyond the Heliopause at about 230 AU is the bow shock which is a plasma wake left by the Sun as it travels through the Milky Way Galaxy.

3.8.12.4 Oort Cloud

The Oort Cloud is a spherical cloud comprising a vast amount of icy particles. It surrounds the Solar System from 50 000 AU to about 100 000 AU from the Sun. The outer limit of the Oort Cloud defines the cosmological boundary of the Solar System and the limit of the Sun's gravitational effect.

The Oort Cloud consists of two regions with a disc shaped inner cloud and a spherical outer cloud. Objects within the cloud mainly comprise frozen water, ammonia, methane, ethane, carbon monoxide and hydrogen cyanide. The cloud is considered to be the remains of the original protoplanetary disc surrounding the Sun. These objects initially coalesced much closer to the Sun but they were drawn out into extremely long elliptic or parabolic orbits by the young gas giant planets. A substantial amount of Oort Cloud comets appear to be from the protoplanetary discs of other stars, being captured during early formation.

3.0 THE UNIVERSE

The spherical shape of outer Oort Cloud is caused by the gravitational influence of nearby stars and galactic tides. The inner, disc shaped cloud has been caused by the Sun.

3.9 Implications for Space Travel

Quantum leaps forward in technological advancement and human genetic engineering are required for successful space travel. For the indefinite survival of the human race, space travel is essential. The only other option is extinction. Distance is the overriding factor in space travel as shown in Table 3.

Table 3. Distances of Solar System Bodies from Earth

Solar System Body	Shortest Distance From Earth (km)
Sun	149 597 871
Mercury	91 690 000
Venus	41 400 000
Moon	384 400
Mars	78 320 000
Jupiter	628 970 000
Saturn	1 283 900 000
Uranus	2 722 900 000
Neptune	4 345 500 000
Pluto	5 720 400 000

Steps towards space travel have commenced and the speed of advancements will be related to progress in science and technology. Successful landings on the Moon have been completed. Skylab was a space station launched by the United States and it orbited Earth from 1973 to 1979. The station was damaged during the launch but numerous scientific experiments were successfully completed during its orbit, prior to crashing back to Earth. Four Salyut space stations have been launched since 1971 with another three more being launched during the development of the Mir Space Station. Mir operated in low Earth orbit from 1986 to 2001, when it crashed to Earth after orbit decay. Mir was the first continuously inhabited space station and significant scientific contribution was made to extraterrestrial effects on humans.

The International Space Station was launched into low Earth orbit in 1998. It is a joint venture project with five participating agencies from the United States, Russia, Japan, Europe and Canada. The International Space Station is the ninth space station and continues to operate in 2016. Much space environment research has been completed as well as numerous scientific experiments. The intention is to replace this station with another orbital space station.

The main problem with orbiting space stations is a temporary life span in which orbital decay eventually results in the stations crashing back to Earth. The immediate future aim should be the establishment of a permanent space station on the Moon, which is 384 400 km from Earth. Such a station would facilitate ongoing human habitation on the Moon and assist with initial missions to Mars.

Lunar surface exploration is required to attempt to locate water sources and usable materials for human survival. The lunar atmosphere is almost a vacuum and cannot be enhanced because the Moon does not have a proper magnetosphere which would protect the atmosphere from solar winds. Lunar space stations would need to be self contained and initially need to be resupplied from Earth.

3.0 THE UNIVERSE

Mars has a number of advantages for human colonisation. It is a minimum distance of 78 320 000 km from Earth. There are two polar ice caps which contain vast quantities of water ice. Liquid water cannot occur on the Martian surface because of low atmospheric pressure. The average temperature is -65°C. Local construction materials are available on the surface of Mars. Advances in chemosynthesis are important for food production on Mars. In this process carbohydrates are made from carbon dioxide and chemical nutrients which are the energy source. Oxidation of inorganic molecules such as hydrogen gas, hydrogen sulphide or methane are the source of energy rather than sunlight in photosynthesis where chlorophyll is required to produce carbohydrates.

A significant problem with the Martian atmosphere is that Mars lost its magnetosphere about 4 billion years ago. Consequently solar winds strip away atoms from the outer layer. The atmosphere consists mainly of carbon dioxide with minor argon and nitrogen and traces of oxygen and water. Solar winds prevent atmospheric enhancement. The surface temperatures vary from about -140°C at the winter polar caps to 35°C in equatorial summer. Because of significant eccentricity of its orbit, the Martian seasons in the southern hemisphere are more extreme and northern seasons are milder than Earth's seasons, even though both planets have a similar axial tilt.

Mars has the largest dust storms of any planet in the Solar System. Such events can range from local to huge storms that cover the entire planet. They result from variations in temperature across the surface and cause strong winds which tend to occur when Mars is closest to the Sun.

Of all the objects in the Solar System, Mars has the most benign conditions for human habitation. There are

AN INTRODUCTION TO THE NEW WORLD ORDER

technical solutions for the physical problems which will be encountered.

Following the successful colonisation of Mars the human race needs to travel further out into the Solar System. Promising targets include Ceres, Ganymede, Europa, Io, Titan and Enceladus.

Ceres is the largest object in the Asteroid Belt which is between Mars and Jupiter (Table 3) and it has a diameter of 940 km. It consists of rock and ice. Water is being ejected from its surface and there may be a buried sea which could be a source of life. Ceres is behaving like a comet with ice patches that can survive for a long time. It could be producing comet-like emissions which are driven by a weak cryovolcano that erupts volatiles such as water, ammonia or methane instead of lava. Where there is a potential for liquid water there is a potential for life.

Ganymede orbits Jupiter which has a closest distance to Earth of 628 970 000 km. It is the largest moon in the Solar System. It has a diameter of 5262 km and is the only moon to have a magnetosphere which provides protection from solar winds. Jupiter's much larger magnetic field also has an influence. Ganymede is composed of about equal amounts of silicate rock and water ice. There is an iron-rich liquid core and a vast internal ocean. A very thin oxygen atmosphere exists. The surface has two types of terrain. The first type consists of dark regions which are characterised by impact craters formed about 4 billion years ago. The remaining two thirds of the surface consists of ridges and grooves which may be a horst and graben structure. Ganymede has polar caps consisting of water ice. A large salt-water ocean affects Ganymede's magnetic field and its aurora. Ganymede probably formed by accretion of gas and dust surrounding Jupiter after the planet formed.

Europa is the second closest of Jupiter's Galilean moons and is the sixth largest moon in the Solar System. It is slightly larger than the Moon with a diameter of 3100 km. It probably has an iron and nickel core surrounded by silicate rock and a water ice crust. Europa has a shiny surface which is probably water ice. The surface is cracked and crumpled. It probably formed by the movement of an underneath liquid ocean. Tidal forces are considered to generate the heat that keeps the ocean liquid. Europa's surface temperature averages -160°C at the equator and -220°C at the poles.

The generation of large planetary tidal waves producing significant kinetic energy, could be the principal heat source of Europa's ocean. The surface has large water vapour plumes when Europa is at its farthest distance from Jupiter. These are caused by significant tidal forces.

Europa has a very thin oxygen atmosphere and a tenuous ionosphere. It also has an induced magnetic field caused by the interaction with Jupiter's magnetic field and the presence of a subsurface conductive layer which is probably a salty, liquid water ocean.

It is distinctly possible that life exists on Europa. Such life could exist in the ocean below the ice in an environment similar to life which exists next to hydrothermal vents along Earth's mid-ocean ridges. At such sites chemosynthetic bacteria and single-celled microorganisms termed archaea form the base of the food chain for organisms such as tube worms, clams, limpets and shrimp. Alternatively life could cling to the lower surface of Europa's ice layers like algae and bacteria in Earth's polar regions. Indications are that cosmic rays which strike Europa's surface convert water ice to oxygen which could be absorbed into the underlying ocean and could support anaerobic and aerobic life. There could also be a significant quantity of hydrogen peroxide on the moon's surface which could provide life sustaining oxygen for the liquid water ocean under the ice crust.

Io is the inner most of Jupiter's Galilean moons and is similar in size to Earth's moon with a diameter of 3643 km. It has an inhospitable environment and is the driest known object in the Solar System. With over 400 active volcanoes including Loki, the most powerful volcano in the Solar System, Io is the most geologically active site in this system. Io has abundant internal heat because it is located between Jupiter's gravity and the gravity of the other large satellites. Tidal forces generate the energy required to maintain constant volcanic activity. Volcanic plumes and lava flows cause significant surface changes.

Jupiter's magnetosphere strips gas and dust from Io's thin atmosphere which consists mainly of sulphur dioxide with minor amounts of sulphur monoxide, sodium chloride, sulphur and oxygen. The atmosphere has considerable variations in density and temperature. The environment is considered to be toxic.

The surface topography consists of smooth plains, lava flows, tall mountains and some depressions. The colourful surface is the result of deposited volcanic exhalative products such as silicates, sulphur and sulphur dioxide frost. The lack of surface water has been caused by Jupiter's hot temperature influence during early evolution. The interior of Io is composed of a silicate crust and mantle and an iron-rich core.

Titan is the largest moon of Saturn which at its closest distance to Earth is 1 283 900 000 km away. The moon is 5150 km in diameter and is covered by a thick orange haze. Its dense atmosphere consists mainly of nitrogen with minor amounts of hydrogen, methane, ethane and acetylene. The haze is caused by the Sun's ultraviolet light breaking down methane in the upper atmosphere.

3.0 THE UNIVERSE

Titan consists of water ice and rocky material. The rocky centre which is about 3400 km thick, is surrounded by ice layers. The interior may still be hot, consisting of liquid water and ammonia between ice layers. The presence of ammonia allows water to remain liquid at a temperature as low as -97°C. The average temperature on Titan is -178°C. Liquid hydrocarbon lakes occur in the polar regions.

Titan remains with Saturn's magnetosphere for most of its orbit, which protects the atmosphere from the solar wind. The presence of atmospheric methane has the greenhouse effect of increasing surface temperature. The haze causes lower temperatures by reflecting sunlight back into space. Titan's clouds which consist of methane result in liquid methane rain.

The surface of Titan is geologically young with ridges and valleys. Dark areas may be methane or ethane seas. Impact craters are present on the surface. Conditions on Titan may be similar to those on early Earth at a significantly lower temperature. Volcanoes appear to have contributed to increased surface methane. Longitudinal dunes consist sand size particles which are probably hydrocarbon polymers.

Titan appears to have a complex organic chemistry environment with a subsurface liquid ocean. The current atmosphere of Titan may be similar in composition to the early atmosphere of Earth. Sufficient organic matter occurs on Titan to start biochemical evolution as occurred on Earth. Life could exist in Titan's liquid methane lakes.

In the distant future about 5 billion years from now when the Sun becomes a red giant, the temperature on Titan could rise sufficiently for liquid water to form and the surface to become suitable for life.

Enceladus which is the sixth largest moon of Saturn has a diameter of 600 km. The surface is covered with water ice and the average temperature is -198°C. Surface features include many impact craters and terrain which has recently been tectonically disturbed. Such activity is still continuing. In the south cryovolcanoes eject geysers of water vapour, volatiles, sodium chloride crystals and ice into space. There are over 100 geysers which are the source of the material in Saturn's E Ring. Geological activity on Enceladus is triggered by tidal heating.

Enceladus has a significant atmosphere with the source being volcanism, geysers and gases escaping from the moon. The atmosphere is composed mainly of water vapour with minor nitrogen, carbon dioxide and methane.

The internal structure of Enceladus appears to be an inner silicate core and an outer water ice mantle. Current geological activity indicates that radioactive and tidal heating have melted the interior.

Surface features on Enceladus include impact craters, smooth terrain, zones of ridges and extensive linear cracks and scarps. The surface is extensively covered with fresh, clean ice which has high reflectance. Tectonic activity has been pronounced forming rifts and cliffs. Two regions of smooth plains have been detected. The south polar region is covered in blue ice. It has a large subsurface ocean of liquid water with a depth of about 10 km. The north polar region is intensely cratered.

Cryovolcanism, which is where water and other volatiles are erupted, has been observed on Enceladus. Geyser jets which have been imaged in the south polar region appear to be similar to geysers on Earth.

Enceladus appears to have a large liquid water ocean beneath its frozen surface. The ocean lies below a substantial ice shelf to 40 km thick. The ocean is about

10 km deep and appears to be thickest in the south polar region.

Warm areas of Enceladus with temperatures as high as -116°C are caused by tidal and radioactive heating. Tidal heating happens through the tidal friction process where orbital and rotational energy are dissipated as heat in the crust of an object. Ammonia which has been detected in plumes, reduces the freezing point of water.

Enceladus has a potentially habitable environment for microbiological life. It has a liquid water ocean, nitrogen in the form of ammonia and organic molecules which include methane, propane, acetylene and formaldehyde. The presence of an internal, salty ocean with an energy source and organic compounds are conducive for the evolution of organic life.

As the Sun transitions into a red giant in about 5 billion years from now the temperature of the Solar System will rise. The surface of inner planets will be incinerated. The asteroid Ceres and the moons Ganymede, Europa, Titan and Enceladus are in the outer Solar System. Solar heating could melt ice on these objects, making them habitable. Ongoing research is needed to determine where in the outer Solar System can human life exist.

4.0 THE EVOLUTION OF LIFE ON EARTH

4.1 Introduction

The Solar System developed from the collapse of a giant molecular cloud about 4.6 billion years ago. The planet Earth formed around 4.54 billion years ago from an accretion disc which comprised dust, hydrogen, helium and other ionised gases. Initially much of the Earth was molten and the initial atmosphere would have consisted of toxic gases exhaled from intense volcanic activity.

Complex organic molecules necessary for life may have formed in the protoplanetary disc before Earth's formation. Such extraterrestrial molecules may have subsequently been transported to Earth via objects such as comets. Chemical reactions in a high energy environment could have produced self-replicating, organic molecules on Earth about 4 billion years ago.

Life first arose from non-living matter on Earth between 3.8 and 3.5 billion years ago. This was the beginning of organic evolution which is defined as the continuous genetic adaptation of organisms or species to the environment by the integrating agencies of natural selection, hybridisation, breeding and mutation. Evolutionary changes result in diversity at every level of life including species, individual organisms and at the molecular level. The present patterns of biodiversity are the result of speciation and extinction.

4.2 Features of Evolution

There are currently about 12 million species of life on Earth which are the result of evolution over 4 billion years. Genetic traits are transmitted from parents to progeny through heredity. Deoxyribonucleic acid

4.0 THE EVOLUTION OF LIFE ON EARTH

(DNA) is the molecule that encodes the genetic instructions which are used in the development and function of living organisms. DNA molecules comprise two biopolymer strands coiled around each other to form a double helix. A genome is an organism's complete set of genetic instructions. Mutations are permanent changes to the DNA sequence of a cell's genome.

The exchange of genes between populations and between species is known as gene flow. The transfer of genes between species results in hybrids.

Natural selection is the process where traits that enhance survival and reproduction become more dominant in successive generations of a population. Natural selection results in evolutionary adaptation to the environment particularly in features such as behaviour and physical appearance.

Adaptation is the alteration in the structure or function of organisms which allows them to survive and multiply in a changing environment. Adaptation is a gradual process which results from natural selection.

Speciation is the natural process where a species diverges into two or more descendant species. A species is the largest group of organisms capable of interbreeding and producing offspring.

Throughout the history of life there have been periods of explosive evolution and periods of mass extinction. Explosive evolution is the diversification of a group of organisms in a short geological time frame.

Index fossils are indicator fossils and are used to identify and define geological periods. Important index fossils include ammonites, corals, graptolites, brachiopods, trilobites and echinoids. A number of microscopic fossils are also useful index fossils.

Extinction is the disappearance of an entire species and it is not unusual. Occasional mass extinctions which have occurred are the result of cataclysmic events.

4.3   Geological Time Scale

The general geological time scale as shown in Table 4 is the systematic, chronological measurement of geological and biological events which have occurred during the history of the Earth. The largest division of geological time is the eon which is divided into eras and then subdivided into periods, epochs and ages.

Geological units from the same time but from different parts of the World often look different and contain different fossils so that the same period was given different names in different locations. The current official geological time scale has been compiled by the International Commission on Stratigraphy.

4.4   Hadean

The Hadean was the first geological eon on Earth and extended from 4.54 billion years ago to 4 billion years ago. This was a violent period for Earth with intense volcanism, frequent collisions with extraterrestrial bodies, toxic outgassing and a largely molten planet.

About 4.5 billion years ago a planet termed Theia about the size of Mars appears to have collided with Earth and blasted material into orbit where it accreted to form the Moon. The impact created a rock vapour atmosphere which would have condensed in several thousands of years leaving behind hot volatiles that resulted in an atmosphere mainly comprising carbon dioxide, hydrogen, helium and water vapour. Liquid water oceans existed despite a surface temperature in excess of 200°C, due to the mainly carbon dioxide atmosphere.

4.0 THE EVOLUTION OF LIFE ON EARTH

Table 4. General Geological Time Scale

Era	Period (Eon)	Epoch	Distinctive Features	Approximate Age (Ma*)
CENOZOIC	Quaternary	Recent	Homo sapiens (modern man)	— 0.011
		Pleistocene	Early man, northern glaciation	— 2.0
	Tertiary	Pilocene	Large carnivores	— 5.0
		Miocene	First abundant grazing mammals	— 25
		Oligocene	Large running mammals	— 34
		Eocene	Modern types of mammals	— 55
		Paleocene	First placental mammals	— 65
MESOZOIC	Cretaceous		First flowering plants, climax of dinosaurs and ammonites, followed by Cretaceous-Tertiary extinction	— 140
	Jurassic		First birds, first mammals, dinosaurs and ammonites abundant	— 210
	Triassic		First dinosaurs, abundant cycads and conifers	— 250
PALEOZOIC	Permian		Extinction of most kinds of marine animals, including trilobites, southern glaciation	— 290
	Carboniferous, Pennsylvanian Sub-period		Great coal forests, conifers, first reptiles	— 330
	Carboniferous, Mississippian Sub-period		Sharks and amphibians abundant. Large and numerous scale trees and seed ferns	— 360
	Devonian		First amphibians, ammonites, abundant fish	— 410
	Silurian		First terrestial plants and animals	— 440
	Ordovician		First fishes, invertebrates dominant	— 505
	Cambrian		First abundant record of marine life, trilobites dominant	— 541
PRE-CAMBRIAN	Proterozoic (Eon)		Fossils extremely rare, protozoa, multi-celled eukaryotes, sponges, jellyfish, primitive aquatic plants, green algae, glaciation	— 2500
	Archean (Eon)		Oldest dated algae, prokaryotes, stromatolites, oxygen producing bacteria	— 4000
	Hadean (Eon)		Organic molecules, RNA, simple cells	
			Birth of Earth	4540

* Ma means millions of years ago

AN INTRODUCTION TO THE NEW WORLD ORDER

The World's oldest dated rock is a 4.4 billion years old zircon crystal from Western Australia's Jack Hills region. This discovery has demonstrated that the Earth's crust appeared soon after the planet formed. This new discovery has also shown that due to the zircon's oxygen isotope composition there already was water on the surface of the Earth at that time. This means that water could have existed on the Earth's surface about 140 million years after its formation. The oldest rock ever dated was a non-terrestrial meteorite which is believed to be 4.56 billion years old.

The period of most intense meteorite impacts occurred during the Late Heavy Bombardment which began about 4.2 billion years ago and finished in the Eoarchean (4 to 3.6 billion years ago). Life could have begun during the Hadean, surviving the Late Heavy Bombardment period in hydrothermal vents below the Earth's surface. Table 5 lists the distinctive features of eras within the Hadean time scale.

Table 5.   Hadean Geological Time Scale

Era	Distinctive Features	Approximate Age (Ma)
Early Imbrian	Indirect photosynthetic evidence of primordial life, self-replicating RNA molecules evolve about 4000 Ma, oldest dated rock 4030 Ma, orogeny in Antarctica	4000
		4100
Nectarian	Named after period of large impact events on the Moon, start of Late Heavy Bombardment	
		4300
Basin Groups	Oldest known mineral (zircon 4400 Ma)	
		4500
Cryptic	Formation of Earth 4540 Ma, formation of the Moon from a giant impact 4530 Ma	
		4540

Considerable changes have occurred to Earth's atmosphere over time. The initial atmosphere which formed from the solar nebula consisted mainly of

4.0 THE EVOLUTION OF LIFE ON EARTH

hydrogen and helium which were dissipated by solar wind and the Earth's heat. The second phase of atmosphere formation resulted in the release of volatile gases from the molten Earth and from impact degassing of incoming extraterrestrial bodies which vapourised on impact. This atmosphere probably contained water vapour, carbon dioxide, nitrogen and minor amounts of other gases such as methane, ammonia and hydrogen sulphide. A third phase of atmosphere formation occurred when bacteria commenced to produce oxygen about 2.8 billion years ago resulting in an atmosphere rich in oxygen. The current atmosphere consists of 78% nitrogen, 21% oxygen and 1% of other gases.

Much of Earth's water is considered to have been supplied by meteorites, comets and other extraterrestrial bodies. As the Earth cooled, clouds formed and rainfall filled oceans which began forming about 4.4 billion years ago and covered the Earth by the end of the Hadean. The presence of greenhouse gases prevented the ocean from freezing.

Conditions for the evolution of life prevailed in the early atmosphere and ocean. Activity from hydrothermal vents may have been conducive to the emergence of life from non-living chemicals. Hydrothermal vents which currently occur along mid-ocean ridges are biologically productive, producing chemosynthetic bacteria and archaea which form the base of the food chain for diverse marine organisms.

Ribonucleic Acid (RNA) is a polymeric molecule which is considered to be the precursor to the start of life on Earth. RNA is assembled as a chain of nucleotides and its various biological roles include coding, decoding, regulation and expression of genes. Self-replicating RNA molecules have been artificially produced in laboratories. In 2015 DNA and RNA organic compounds including uracil, cytosine and thymine were created in a laboratory under extraterrestrial conditions using starting chemicals such as pyrimidine, found in meteorites. Pyrimidine is an aromatic, heterocyclic, organic compound which is the most carbon-rich chemical found in the Universe.

The formation of double-walled bubbles of lipids or liposomes may have been an initial step to the evolution of external membranes of cells. Nucleic acids such as RNA could have more easily formed inside of liposomes.

4.5   Archean

The Archean eon extended from 4 billion years ago to 2.5 billion years ago. The divisions are listed in Table 6.

Table 6.   Archean Geological Time Scale

Era	Distinctive Features	Approximate Age (Ma)
Neoarchean	Stabilisation of most modern cratons, orogenies, greenstone belts	2500
Mesoarchean	First stromatolites (cyanobacteria), oldest macrofossils	– 2800
Paleoarchean	First known oxygen-producing bacteria, oldest definitive microfossils, oldest cratons on Earth	– 3200
Eoarchean	Simple single-celled life comprising bacteria and archaea, oldest probable microfossils, end of Late Heavy Bombardment	3600
		4000

Archean heat flow was significantly higher than today due to a mixture of left over heat from planetary accretion, heat from the formation of the Earth's core and heat generated by the decay of radioactive elements. Archean rocks are predominantly metamorphic or igneous. Volcanic activity was intense and common rock types include extrusive komatiites and high-magnesium basalts, large granite plutons and metamorphic gneisses. The oldest surviving rocks are generally not older than

4.0 THE EVOLUTION OF LIFE ON EARTH

3.8 billion years because in the period from 3.8 billion years to 4.2 billion years ago the Earth's crust was pulverised during the Late Heavy Bombardment.

The Archean atmosphere did not have free oxygen and the presence of greenhouse gases contributed to higher ambient temperatures.

The first large pieces of continental crust appeared at the beginning of the Archean about 4 billion years ago. These cratons were the cores around which continents accreted. They comprised greenstone belts which formed in subduction zones and felsic, magmatic, granitic rocks which formed the first continental crust. By the end of the Archean plate tectonic activity was similar to that of the modern Earth. From this period 2.5 billion years ago there are well-preserved sedimentary basins, volcanic arcs, intercontinental rifts and widespread orogenic events which indicate the accretion and disintegration of several supercontinents. Orogeny is the large structural deformation of the Earth's lithosphere resulting from interaction between tectonic plates. Orogenic belts are formed when tectonic plates collide and mountain ranges are formed.

Liquid water was common and deep ocean basins were present as indicated by banded iron formations, chert beds, chemical sediments and pillow basalts. Only a fraction of the current amount of continental crust formed in the Archean.

According to scientific evidence life on Earth began between 3.5 and 3.8 billion years ago. Biogenic graphite was found in 3.7 billion years old metasedimentary rocks from Western Greenland. Stromatolite fossils were discovered in 3.48 billion years old sandstone from Western Australia.

Initial life in the Archean probably occurred as simple non-nucleated, single-celled organisms called prokaryotes. The oldest known fossilised prokaryotes are about 3.5 billion years old. A prokaryote lacks a membrane-bound nucleus, mitochondria or any other membrane-bound organelles. All the intracellular components such as proteins, DNA and metabolites are located together and are enclosed by the cell membrane. Some prokaryotes have multicellular stages to their life cycles while others like cyanobacteria form large colonies.

Prokaryote cells can be divided into two domains which are archaea and bacteria. A third domain within the species comprises eukaryotes which evolved about 2 billion years ago. These organisms have a well defined membrane-bound nucleus containing chromosomal DNA and organelles.

Bacteria and archaea reproduce asexually through binary fission. Prokaryotes live in all environments on Earth. Some archaea and bacteria which survive in harsh environments of high temperatures or high salinity, are known as extremophiles and they can obtain energy from inorganic compounds like hydrogen sulphide.

The last universal ancestor (LUA) is the most recent organism from which all present day organisms living on Earth have evolved from. It lived in early Archean eon about 3.5 billion years ago. This ancestor appears to have been a prokaryote which had a cell membrane and ribosomes but lacked a nucleus or membrane-bound organelles such as mitochondria and chloroplasts. It used DNA as the genetic code, RNA for information transfer and protein synthesis and enzymes to catalyse reactions. The cell multiplied by duplicating all its contents followed by cellular division. The cell used chemiosmosis to produce energy and it reduced carbon dioxide and oxidised hydrogen.

Stromatolites, which are laminated, calcareous, sedimentary formations produced by lime-secreting

4.0 THE EVOLUTION OF LIFE ON EARTH

cyanobacteria, evolved about 3.5 billion years ago. Lichen stromatolites are considered to be types of layered rock structure that form above water in the water/rock contact zone.

Stromatolites are abundant in the fossil record and were one of the earliest forms of life on Earth. Their evolution peaked about 1.25 billion years ago and then their abundance declined as they became victims to other animals. One genus of stromatolite which is common in the geological record is Collenia. This genus of fossil cyanobacteria occurred in banded iron formations more than 2 billion years ago.

Cyanobacteria were responsible for increasing the oxygen level of the early atmosphere through photosynthesis. Cyanobacteria use water, carbon dioxide and sunlight to create their food. A layer of mucus forms over mats of cyanobacterial cells and debris from the surrounding environment becomes trapped within the mucus. This debris becomes cemented with calcium carbonate to form thin laminations of limestone.

Modern stromatolites occur in hypersaline environments such as lakes and marine lagoons where conditions exclude predatory grazers.

Towards the end of the Archean eon most modern cratons had stabilised and greenstone belts had been emplaced.

4.6 Proterozoic

4.6.1 Introduction

The Proterozoic eon occurred from 2.5 billion years ago to 541 million years ago. The geological record for this eon is much better preserved than the Archean record. Proterozoic geology includes extensive strata deposited in shallow seas and is typified by much less rock metamorphism and alteration than in the Archean. Cratons accreted into continents of current sizes and orogenic events were common. Life forms developed from prokaryotes to eukaryotes and multicellular organisms. Several severe ice ages occurred in this eon. The Great Oxygenation Event (GOE) was a major environmental event which affected the atmosphere and anaerobic life. Distinctive features of the Proterozoic are listed in Table 7.

4.6.2 Great Oxygenation Event

The oxygen catastrophe occurred in the Siderian period between 2.3 and 2.5 billion years ago. The increase in oxygen was the result of photosynthesis in vast numbers of single-celled algae and cyanobacteria. Although oxygen had been released by photosynthesis as far back as the Archean Eon it was not until the Siderian period that chemical sinks of unoxidised iron, sulphur and limestone were full. The formation of oxidised iron in the form of red layers in banded iron formations was abundant in the Siderian period. Following widespread oxidising reactions, free oxygen began to accumulate in the atmosphere to form Earth's third atmosphere. The rising oxygen levels destroyed a huge portion of Earth's anaerobic organisms including cyanobacteria, which resulted in the largest extinction event in Earth's history. As well the free oxygen combined with methane in the atmosphere causing the Huronian glaciation. Methane is a strong greenhouse gas, but it combines with oxygen to form water and carbon dioxide, which is a less effective greenhouse gas.

Some of the released oxygen was stimulated by incoming ultraviolet radiation to form ozone which concentrated in the upper atmosphere. The ozone layer protects the Earth from incoming ultraviolet radiation.

4.0 THE EVOLUTION OF LIFE ON EARTH

Table 7.   Proterozoic Geological Time Scale

Era	Period	Distinctive Features	Approximate Age (Ma)
Neoproterozoic	Ediacaran	Good fossils of the first multi-celled animals, Ediacaran biota flourish worldwide in seas, simple trace fossils of possible worm-like Trichophycus, first sponges and trilobitomorphs, many soft-jellied creatures, various orogenies	541 — 650
	Cryogenian	'Snowball Earth' period, fossils still rare, Rodinia landmass begins to break up	— 850
	Tonian	Rodinia supercontinent persists, trace fossils of simple multi-celled eukaryotes, first radiation of dinoflagellate-like acritarchs, orogenies, geosynclines	— 1000
Mesoproterozoic	Stenian	Narrow highly metamorphic belts due to orogeny as Rodinia forms, various orogenies	— 1200
	Ectasian	Green algae colonies in the ocean, orogeny in North America	— 1400
	Calymmian	Australian orogenies	— 1600
Paleoproterozoic	Statherian	First complex single-celled life comprising protists with nuclei, Columbia is the primordial supercontinent, Australian orogenies	— 1800
	Orosirian	The atmosphere becomes oxygenated, Vredefort and Sudbury Basin asteroid impacts, much orogeny, Gawler craton in Australia	— 2050
	Rhyacian	Bushveld Igneous Complex forms, Huronian glaciation	— 2300
	Siderian	Oxygen catastrophe, banded iron formations deposited, Australian orogeny	2500

Oxygen was toxic to much life on Earth and resulted in large scale extinction. Some organisms adapted and developed the ability to use oxygen to improve their metabolic rate.

4.6.3   Huronian Galaciation

The Huronian glaciation occurred from 2.4 to 2.1 billion years ago following the Great Oxygenation Event which increased atmospheric oxygen and decreased atmospheric methane. Drawdown of atmospheric carbon dioxide also occurred during this period. The depletion of greenhouse gases resulted in the longest and most severe ice age in geological history, at a time when the Earth was populated by unicellular life.

4.6.4   Columbia Supercontinent

The supercontinent Columbia formed between 2.1 and 1.8 billion years ago. Large scale collisional events resulted in the continental accretion of the cratons of Laurentia, Baltica, Ukrainian and Amazon shields, Australia, Siberia, North China and Kalaharia to form Columbia. A number of subduction zones occurred during these tectonic events.

The break up of Columbia commenced about 1.6 billion years ago with widespread granitic, magmatic activity in North America, Baltica, Amazonia and North China.

4.6.5   Rodinia Supercontinent

The supercontinent Rodinia which existed between 1.1 billion and 750 million years ago, contained nearly all of the Earth's land mass. The accretion comprised the cratons of North America (Laurentia), East Europe, Amazonia, West Africa, Rio Plato, Sao Francisco, Congo, Kalahari, Australia, India, East Antarctica, Siberia and South China.

4.0 THE EVOLUTION OF LIFE ON EARTH

The land mass of Rodinia was centred around the Equator and was barren. The break up of this supercontinent occurred in the Cryogenian period. Rifting progressively occurred over about 100 million years forming new oceans. Volcanic activity associated with the formation of rift valleys caused contamination of the atmosphere with volcanic ash and dust, resulting in glaciation events termed Snowball Earth.

4.6.6 Snowball Earth

Three ice ages occurred in the Neoproterozoic with glaciers extending from the poles to the Equator. These events occurred at 710 million (Sturtian glaciation), 635 million (Marinoan glaciation) and 580 million years ago (Ediacaran Gaskiers glaciation). As the Earth became frozen the white surface reflected sunlight, further reducing the temperature and extending the ice cover. Temperature on Snowball Earth reached minus 50°C.

The three ice ages coincided with the break up of the supercontinent Rodinia. The formation of rift valleys and associated volcanism could have been responsible for the glaciation events by contaminating the atmosphere.

The cessation of global glaciation probably occurred by the release of greenhouse gases such as carbon dioxide and methane from volcanic activity along rift valleys. This allowed the atmosphere to heat up and catalysed a population explosion of cyanobacteria, resulting in atmospheric oxygenation. Such conditions contributed to the Cambrian evolutionary explosion.

4.6.7 Proterozoic Life

Even though the Proterozoic eon extended over 1959 million years the fossil record of organisms is exceedingly sparse. Although the physical conditions were not as harsh as in the Hadean and Archean eons, evolution was inhibited. Causes included major glaciations, atmospheric changes including the Great Oxygenation Event, ongoing orogenic events and significant plate tectonics activity including the formation and break up of supercontinents with associated volcanic activity. By the end of the Proterozoic there was a stable, oxygenated atmosphere and extensive shallow seas, which set up conditions for Cambrian explosive evolution.

Prokaryotes comprising bacteria and archaea which evolved in the Archean also existed in the early Proterozoic. Eukaryotes which evolved about 2.1 billion years ago are different to prokaryotes in that in an eukaryote the cells contain a nucleus and other organelles such as mitochondria and chloroplasts, enclosed within membranes. Many unicellular organisms such as protozoa and all multicellular organisms are eukaryotes including animals, plants and fungi. Asexual prokaryotes and eukaryotic phytoplankton which were the prevalent life forms in the early Proterozoic reproduced by mitosis where one cell divides to produce two genetically identical cells. Sexually reproducing eukaryotes reproduce by meiosis in which the maturation process of gametes consists of chromosome conjugation and two cell divisions in the course of which the diploid chromosome number becomes reduced to the haploid. This allows the gametes to fuse through fertilisation to form a zygote which is a mixture of maternal and paternal chromosomes.

The domain Eukaryota is monophyletic and forms one of the three domains of life with the other two being the prokaryotes bacteria and archaea. Acritarchs are ubiquitous, kerogen fossils which appeared about 2.1 billion years ago and are considered to be eukaryotes. The unicellular fossils of eukaryotes are more complex with external morphological projections. Older acritarchs are probably related to unicellular marine algae. Acritarchs include the remains of organisms including the egg cases of small metazoans to cysts of green algae. Red and green algae appeared over 1 billion years ago.

Stromatolites evolved about 3.5 billion years ago. Stromatolite diversity continued to expand throughout the Proterozoic until about 1 billion years ago. Herbivorous eukaryotes began feeding on stromatolites about 700 million years ago and the population progressively declined.

Grypania is a tube like fossil from near the Paleoproterozoic-Mesoproterozoic boundary. It may have been a bacterial colony or eukaryotic algae.

Sexually reproducing phytoplankton appeared in large numbers in the later Proterozoic. From about 650 to 541 million years ago which is known as the Vendian period, the first diversification of soft-bodied organisms known as Vendian fauna occurred. Although many of the fossils are difficult to interpret they include lichens, large protozoans, cnidarians, worms, sponges, soft-jellied creatures and soft-bodied relatives of arthropods. Vendian rocks also contain trace fossils which are indirect evidence of life. They include tracks, trails, burrows and impressions made by various invertebrates.

4.7 Paleozoic

4.7.1 Introduction

The Phanerozoic which is the current eon on Earth started 541 million years ago. It comprises three eras which are the Paleozoic (541 to 251 Ma), Mesozoic (251 to 65 Ma) and Cenozoic (65 Ma to present).

The Paleozoic era was a time of major geological, climatic and evolutionary change with the Cambrian period having the most rapid and extensive diversification of life in Earth's history. Most modern phyla first appeared in the Cambrian and fish, arthropods, amphibians and reptiles all evolved during the Paleozoic. Life migrated from the oceans to land and great forests of primitive plants formed large coal deposits. Towards the end of the era large reptiles were prevalent and the first modern plants appeared in the form of conifers.

The Paleozoic era ended with the Permian-Triassic extinction event which was the largest mass extinction in Earth's history resulting in the death of 95% of all marine species.

The Paleozoic geological time scale with distinctive features is summarised in Table 8.

4.7.2 Cambrian

The Cambrian period which extended from 541 Ma to 485 Ma, marked a rapid diversification of life forms in what is known as the Cambrian explosion. This event produced the first representatives of all modern animal phyla. Numerous organisms evolved in the oceans. The land was barren, being dry, rocky and lacking vegetation.

During this time the supercontinent Rodinia broke up to form Pangaea. Gondwana formed prior to Pangaea and then became part of Pangaea which comprised Laurasia and Gondwana. The land masses which formed Gondwana were Antarctica, South America, Africa, Madagascar, Australia, the Arabian Peninsula and India.

The Earth was cold during the early Cambrian with polar ice caps and a series of glaciations. Average temperature was 21°C. Atmospheric carbon dioxide levels were 20 to 35 times current levels (6000 ppm compared to today's 350 ppm). Atmospheric oxygen content was 12.5%. The sea level rose steadily to be 30 m to 90 m above current sea level.

4.0 THE EVOLUTION OF LIFE ON EARTH

Table 8. Paleozoic Geological Time Scale

Period	Epoch	Distinctive Features	Approximate Age (Ma)
Permian	Copingian	Permian-Triassic extinction eliminating 95% of marine species including all trilobites, graptolites and blastoids and 70% of terrestrial species, land masses form supercontinent Pangea, synapsid reptiles, coal age flora, seed plants and conifers diversify, beetles and flies evolve, productid and spiriferid brachiopods, pelecypods, forams, ammonites, terrestrial temnospondyl amphibians and pelycosaurs, orogenies in North America, Europe, Asia and Australia, Permo-Carboniferous glaciation	251
	Guadalupian		260
	Cisuralian		271
			299
Carboniferous	Pennsylvanian	Winged insects, large amphibians, first reptiles, coal forests, highest ever atmospheric oxygen levels, goniatites, brachiopods, bryozoa, pelecypods, corals, forams, orogenies in Europe and Asia	
			323
	Mississippian	First amniote vertebrates, synapsids, large primitive trees, first land vertebrates, rhizodonts, first large sharks, ratfish and hagfish, echinoderms, corals, bryozoa, goniatites, brachiopods, trilobites and nautiloids in decline, glaciation in East Gondwana	
			359
Devonian	Late	Late Devonian extinction of 70% of all invertebrates caused by widespread ocean anoxia (lack of oxygen), and depletion of CO_2 caused by continental greening, first club mosses, ferns, seed bearing plants and trees, land flora dominated by seed ferns, first insects and crabs, sharks, tetrapods brachiopods, corals, ammonoids, coleoids, trilobites, ray finned fish, amphibians, mites, first lichens and stoneworts, orogenies in North Africa, North America and New Zealand	382
	Middle		393
	Early		
			419
Silurian	Pridoli	First vascular plants, millipedes, arthropleurids, trigonotarbid arachnids, land scorpions, jawed and agnathan fish, armoured jawless fish, sea scorpions, corals, brachiopods, trilobites, molluscs and graptolites, first primitive land plants evolved from green algae accompanied by fungi, Caledonian Orogeny	423
	Ludlow / Cayugan		427
	Wenlock		431
	Llandovery / Alexandrian		
			443
Ordovician	Late	Major Ordovician-Silurian extinction event resulting in the death of 60% of marine invertebrates caused by glaciation, diversification of invertebrates, early corals, brachiopods, pelecypods, nautiloids, trilobites, ostracods, bryozoa, echinoids, conodonts, first vertebrates with true bones (jawless fish)	458
	Middle		470
	Early		
			485
Cambrian	Furongian	Cambrian explosion of life in the ocean with appearance of chordates, arthropods (trilobites and crustaceans), echinoids, brachiopods, cephalopods, chitons, sponges, graptolites, priapulid worms, archaeocyatha, anomalocarids, foraminifera, radiolaria, prokaryotes, fungi and algae, emergence of Gondwana orogenies in Australia and Antarctica, atmospheric carbon dioxide levels were 20 to 35 times current levels	497
	Series 3		509
	Series 2		521
	Terreneuvian		541

Although no true land plants are known from the Cambrian, biofilms and microbial mats developed in tidal areas. Lichen, fungi and microbes occurred on land.

Most of the animal life in the Cambrian was marine with trilobites being dominant. Organisms which evolved included prokaryotes, radiolaria, foraminifera, anomalocarids, archaeocyatha (a sponge which became extinct in the Cambrian), other sponges, priapulid worms, graptolites, chitons, cephalopods, brachiopods, echinoids, arthropods (trilobites, crustaceans) and chordates.

Diversification which occurred early in the Cambrian changed later in the period due to oxygen levels falling significantly in the oceans, leading to hypoxia while the levels of toxic hydrogen sulphide rose.

4.7.3   Ordovician

The Ordovician period spanned from 485 Ma to 443 Ma and life continued to evolve although there was a mass extinction towards the end. Invertebrates such as molluscs and arthropods were dominant in the oceans and fish continued to evolve. Molluscs included pelecypods, gastropods and nautiloid cephalopods. Graptolites thrived in the oceans and the first true vertebrates being fish (ostracoderms) evolved. The first jawed fish (gnathostomes) appeared in the late Ordovician.

New types of organisms included tabulate corals, brachiopods, graptolites, conodonts and echinoderms. Green algae were common and the first land fungi evolved.

During the Ordovician the southern continents which were combined to form Gondwana drifted towards the South Pole. During the middle Ordovician there was a meteor shower event. In the beginning of the late

Ordovician intense volcanic activity released large quantities of carbon dioxide, increasing the atmospheric temperature. The mean surface temperature was 16°C with an atmospheric oxygen content of 13.5% and a carbon dioxide level of 4200 ppm. The sea level was 180 m above present day rising to 220 m and then falling sharply to 140 m at the end of the Ordovician glaciations.

The Ordovician-Silurian extinction events were caused by a fall in the atmospheric carbon dioxide level from 7000 ppm to 4000 ppm resulting in widespread glaciations with ice caps forming on Gondwana. About 60% of marine invertebrates became extinct including conodonts, graptolites, some groups of trilobites, brachiopods and bryozoans.

An alternative extinction possibility is that a gamma-ray burst destroyed the ozone layer, exposing Earth's life to deadly radiation and initiating global cooling.

4.7.4 Silurian

From 443 Ma to 419 Ma the geological period is termed the Silurian. Evolutionary advances included the diversification of jawed and bony fish, the appearance on land of the first vascular plants and the arrival of small terrestrial arthropods.

During the Silurian Gondwana continued to drift southward. Europe collided with North America, this event being the Caledonian orogeny. A vast ocean covered much of the northern hemisphere. The sea level was about 180 m above present day. The Earth entered a long greenhouse phase with warm shallow seas. The mean surface temperature was 17°C. Atmospheric oxygen and carbon dioxide levels were 14% and 4500 ppm respectively.

Vascular plants appeared in the second half of the Silurian period. Plants included moss forests, Cooksonia (in the northern hemisphere), Baragwanathia (in Australia) and Psilophyton. Corals, graptolites, molluscs, trilobites and sea scorpions continued to evolve. Armoured jawless fish and jawed and agnathan fish appeared. The earliest known terrestrial animals which evolved in the mid Silurian included millipedes, centipedes, arthropleurids, trigonotarbid arachnids and land scorpions. The first known air-breathing animals were arthropods.

4.7.5 Devonian

During the Devonian period which spanned from 419 Ma to 359 Ma vascular plants formed extensive forests, arthropods became established and fish species showed considerable diversification. This time zone is called the "Age of Fish". The first ray finned and lobe finned bony fish appeared and placoderms (armoured fish) were prevalent. The ancestors of all tetrapods began adapting to walking on land by using strong pectoral and pelvic fins which subsequently evolved into legs. Life in the oceans consisted of primitive sharks, ammonites, trilobites, brachiopods and corals.

The supercontinent Gondwana occupied most of the southern hemisphere and began a significant northerly drift. The continent Euramerica (or Laurussia) formed early in the Devonian by the collision of Laurentia and Baltica. Considerable tectonic activity occurred as Euramerica and Gondwana approached each other. The Caledonian Orogeny continued.

Sea level was high being 189 m above present level falling to 120 m through the period. The mean surface temperature was 20°C and atmospheric oxygen and carbon dioxide levels were at 15% and 2000 ppm respectively.

Extensive reef building occurred in shallow water at this time. The reefs comprised calcareous algae,

4.0 THE EVOLUTION OF LIFE ON EARTH

stromatoporoids and tabulate and rugose corals. Marine faunas included bryozoa, corals, brachiopods, ammonites, crinoids, ostracoderms (armoured fish), gnathostomes (jawed fish) and armoured placoderms. Cartilaginous and bony fish became diverse with sharks being prevalent.

Land animals such as millipedes, centipedes and arachnids continued to diversify. Early tetrapods probably evolved from lobe finned fish.

Plants which began colonising the land during the Silurian period continued to diversify. Lycophytes, horsetails and ferns grew to large sizes and formed the first forests. By the end of the Devonian progymnosperms such as Archaeopteris were the first trees to evolve. The proliferation of plants increased the oxygen content of the atmosphere while simultaneously reducing the level of the greenhouse gas, carbon dioxide, facilitating the extinction event.

The late Devonian extinction resulted in the death of 70% of all invertebrates while terrestrial plants and animals were largely unaffected. The causes of this catastrophic event were depletion of atmospheric carbon dioxide caused by continental greening and widespread ocean anoxia (lack of oxygen). The most severely affected species were the brachiopods, trilobites, ammonites, corals, conodonts, acritarchs, jawless fish and all placoderms.

The Devonian extinction comprised two prolonged episodes of species depletion and several shorter periods. The Kellwasser Event in late middle Devonian resulted in the obliteration of the great coral reefs, jawless fish and the trilobites. The Hangenberg Event at the boundary of the Devonian and Carboniferous destroyed the placoderms and most early ammonites.

4.7.6 Carboniferous

The Carboniferous period which extended from 359 Ma to 299 Ma, is divided into the Mississippian epoch (359 Ma to 323 Ma) and the Pennsylvanian epoch (323 Ma to 299 Ma). Large thicknesses of coal bearing deposits were laid down during the Pennsylvanian epoch in Europe and North America. Temperatures were high in the early Carboniferous averaging 20°C with tropical swamps being prevalent. During the middle Carboniferous temperatures reduced to 10°C with the average temperature for the period being 14°C. The atmospheric content of oxygen attained the highest level in history, being 35% compared with 21% today. Average atmospheric carbon dioxide level was 800 ppm. The sea level fell from 120 m above present day to present day level during the Mississippian and then rose to about 80 m above present level at the end of the period.

The most important evolutionary event in the Carboniferous period was the development of amniotic eggs, which allowed amphibians, the dominant existing vertebrates, to migrate further inland. The first reptiles and synapsids evolved in swamps. The atmospheric temperature cooled during the Carboniferous leading to the glaciation of Gondwana in an event known as the Permo-Carboniferous glaciation.

Active mountain building occurred in the Carboniferous as the supercontinent Pangaea came together. Gondwana remained intact but collided with North America and Europe (Laurussia) resulting in the Hercynian orogeny in Europe and the Alleghenian orogeny in North America.

Large coal deposits of the Carboniferous resulted from the appearance of bark bearing trees containing lignin and lower sea levels, resulting in extensive swamps and forests in Europe and North America.

4.0 THE EVOLUTION OF LIFE ON EARTH

The main early Carboniferous plants were horsetails, vines, club mosses, scale trees, ferns and seed ferns. Cycads and conifers appeared. Large trees with trunks to 30 m were common. Conifers evolved later in the Carboniferous, adapting to higher and drier ground. All modern classes of fungi were present in the late Carboniferous.

Shallow, warm marine waters often flooded the continents. Prevalent marine invertebrates included forams, corals, bryozoa, ostracods, molluscs and echinoderms. Annelids were common and pelecypods continued to expand. Trilobites were trending towards extinction, but long stemmed crinoids flourished. Fresh water invertebrates included pelecypods and eurypterids.

Terrestrial invertebrates included insects, myriapods, arachnids, priapulids, arthropods such as Arthropleura, griffin flies, dragon and may type flies, and ancestors of cockroaches. Land snails also evolved.

Fish were prevalent in the Carboniferous oceans and included elasmobranchs (sharks and their relatives), bony fish and freshwater fish. Carboniferous amphibians were abundant and reached lengths of 6 m. The Permo-Carboniferous glaciation adversely affected amphibian evolution.

Reptiles had a major evolutionary expansion. By the end of the Carboniferous period they had diversified into a number of groups including protorothyridids, captorhinids, araeoscelids and pelycosaurs.

The Mississippian stratigraphy comprises shallow water limestone. Also present are sandstone and siltstone. Coal seams to 12 m thick are prevalent in the late Carboniferous. Multiple transgressions and regressions of the sea occurred in the Pennsylvanian resulting in sedimentation layers varying from sandstone, siltstone, shale, freshwater limestone, underclay and coal. Changing sea levels saw lithology changes from shale (tidal shoreline) to limestone (shallow marine) and black shale (deep marine).

4.7.7 Permian

The Permian which was the last period of Paleozoic era, extended from 299 Ma to 251 Ma. By the early Permian, Gondwana and Euramerica had collided to form the supercontinent Pangaea and there was one global ocean called Panthalassa. Because of the large size of Pangaea the interior of this continent was much cooler and drier than the climate in the Carboniferous. The extensive Carboniferous rain forests had vanished leaving large areas of arid deserts. Amniotes such as the ancestral groups of the mammals, turtles, lepidosaurs and archosaurs became dominant at the expense of amphibians, due to the drier conditions. At the end of the Permian period the greatest mass extinction in Earth's history occurred with 95% of marine species and 70% of terrestrial species being obliterated. It took ecosystems about 30 million years to recover.

The sea level remained at about 60 m during the early Permian but dropped to -20 m in the late Permian. The mean atmospheric oxygen and carbon dioxide levels were 23% and 900 ppm respectively. The average temperature was 16°C. Throughout the Permian the climate was varied. Early on there was the Permo-Carboniferous glaciation. The climate gradually became hotter and was warm in the late Permian. Orogenies occurred in North America, Europe, Asia and Australia.

Permian marine deposits contain abundant fossil molluscs and echinoderms. Around the Pangaea continental shelf reefs were large with numerous sponge and coral species. Ammonites similar to nautilus and brachiopods were common. Early fish were replaced by true bony fish. Sharks and rays prevailed.

4.0 THE EVOLUTION OF LIFE ON EARTH

On land the giant swamp forests of the Carboniferous began to dry out and moss type plants were replaced by gymnosperms which were the first seed bearing plants. In the Permian there was radiation of many conifer groups. Extensive seed fern forests of Glossopteris flora prevailed in the southern continent. Large deposits of bituminous coal occur within Russian and Australian Permian strata.

Arthropods continued to diversify during the Permian period. Primitive relatives of cockroaches and dragonflies were prevalent. New groups included cicadas and beetles.

Early Permian terrestrial faunas included pelycosaurs, diadectes and amphibians. During the middle Permian therapsids became prevalent. The Permian landscape became dominated by synapsids and sauropsids which included the Dimetrodon. Reptiles became the dominant vertebrates because of their ability to survive in a drier climate. Some amphibians persisted. At the end of the Permian the largest synapsids became extinct. The sauropsids survived the Permian extinction event and gave rise to the dinosaurs that would be dominant in the Mesozoic era.

The Permian ended with the greatest extinction event ever recorded in paleontology. In the Permian-Triassic extinction event 95% of all marine species and 70% of all terrestrial organisms became extinct. Catastrophic volcanic activity in Siberia and China contributed to this event. Massive flood basalt eruptions resulted in lava flows lasting thousands of years, which formed the Siberian Traps. Vast quantities of carbon dioxide were released. Lowered sea levels and volcanic fallout resulted in much higher carbon dioxide levels in the oceans resulting in decimation of marine ecosystems.

Ocean venting of hydrogen sulphide gas may have also occurred resulting in anoxic zones. Atmospheric release of hydrogen sulphide would destroy the ozone layer allowing ultraviolet radiation to penetrate to the Earth's surface, destroying species.

There are indications of a massive asteroid hitting Antarctica at the end of the Paleozoic era. The Permian-Triassic extinction event was probably caused by a number of factors to explain the massive extent of species decimation.

4.8 Mesozoic

4.8.1 Introduction

The Mesozoic era extended from 251 Ma to 65 Ma and is called the age of reptiles. The three periods of this era are the Triassic, Jurassic and Cretaceous. Table 9 is a summary of the Mesozoic geological time scale with distinctive features.

The Mesozoic began with the Permian-Triassic mass extinction event resulting in the greatest decimation of species in Earth's history and ended with the Cretaceous-Paleogene extinction event which killed off dinosaurs.

In the Mesozoic there was considerable tectonic, climatic and evolutionary activity. Pangaea rifted into separate land masses. The climate varied between warm and cool periods, generally being hotter than present day. Dinosaurs became the dominant terrestrial vertebrate until their extinction in the Cretaceous. The first birds and mammals appeared in this era.

4.8.2 Triassic

The Triassic was hot and dry with deserts extending over much of Pangaea until tectonic rifting occurred to form Laurasia and Gondwana. Red bed sandstones and evaporites were common. Average temperature was 17°C. Mean oxygen and carbon dioxide contents were 16% and 1750 ppm (five times above current level), respectively.

4.0 THE EVOLUTION OF LIFE ON EARTH

Table 9. Mesozoic Geological Time Scale

Period	Epoch	Distinctive Features	Approximate Age (Ma)
Cretaceous	Late	Mass extinction at the end of the Cretaceous caused by an asteroid with the loss of 75% of the Earth's species, flowering plants prevalent, new insects, modern fish, ammonites, pelecypods, echinoids and sponges common, many new dinosaurs, modern sharks, primitive birds, marsupials, placental mammals, break up of Pangaea, North American orogenies, extensive chalk deposits, break up of Gondwana, carbon dioxide levels similar to present day level	65
			-- 100
	Early		
			-- 145
Jurassic	Late	Gymnosperms (Bennettitales, conifers, cycads) gingkoes and ferns common, many types of dinosaurs, small mammals, first birds (like Archaeopteryx) and lizards, ichthyosaurs, plesiosaurs, pelecypods, ammonites, belemnites, echinoids, crinoids, brachiopods and sponges common, coralline algae, break up of Pangaea into Laurasia and Gondwana, North American orogeny, carbon dioxide levels to 1950 ppm resulting in higher global temperatures	-- 163
	Middle		-- 175
	Early		
			-- 200
Triassic	Late	Triassic-Jurassic extinction event with death of all archosaurs, synapsids, almost all amphibians and 34% of all marine life, archosaurs dominant on land prior to this event, ichthyosaurs and nothosaurs in the oceans, flying pterosaurs, temnospondyl amphibians, cynodonts, first mammals and crocodiles, ammonites, corals, fish, insects, plants including conifers, lycophytes, cycads, ginkgophytes and glossopterids, break up of Pangaea and formation of the Tethys Sea, orogenies in South America, Asia, New Zealand and Australia, Permian-Triassic extinction event wiping out 95% of marine species and 70% of terrestrial species	-- 237
	Middle		-- 247
	Early		
			251

AN INTRODUCTION TO THE NEW WORLD ORDER

The Permian-Triassic extinction event resulted in the decimation of many species on Earth. Surviving species included Lystrosaurus, Labyrinthodont and Euparkeria. Temnospondyli evolved at this time and became a dominant predator.

During the Triassic nearly all of the Earth's land mass was coalesced in the supercontinent of Pangaea. Africa was joined to Pangaea as was South America. The break up of Pangaea commenced and the Tethys Sea formed. Orogenies occurred in South America, Asia, New Zealand and Australia.

Terrestrial plants of this period included conifers, lycophytes, cycads, ginkgophytes and glossopterids. Glossopteris was a prevalent southern hemisphere seed fern in the early Triassic.

Triassic marine fauna included modern types of corals, serpulids, microconchids, ammonites, fish, and marine reptiles including pachypleurosaurus, nothosaurs, placodonts, plesiosaurs, askeptosaurs and ichthyosaurs. Terrestrial fauna included temnospondyl amphibians and various reptiles which included mastodonsaurus, rhynchosaurs, archosaurs and theropods. Turtles appeared in the late Triassic. Cynodonts which are a large group that includes mammals evolved in the Permian and flourished in the Triassic.

The Triassic-Jurassic extinction event resulted in the death of all archosaurs, synapsids, almost all amphibians and 34% of all marine life. This event was probably caused by massive volcanic eruptions resulting in the flood basalts of the Central Atlantic Magmatic Province, as the supercontinent Pangaea started to break apart. Volcanic activity would have released ash and gases such as carbon dioxide and sulphur dioxide into the atmosphere.

4.8.3 Jurassic

The start of this period was marked by the Triassic-Jurassic extinction event. Pangaea has been rifting into two land masses being Laurasia to the north and Gondwana to the south, separated by the Tethys Sea. Two other extinction events which occurred in this period were in the early Jurassic and at the end. Gondwana began to break up in the early Jurassic accompanied by a flood volcanic event that produced a huge quantity of volcanic rocks ranging from basalts to rhyolites.

In the Jurassic oxygen and carbon dioxide levels were 26% and 1950 ppm respectively. The average temperature was 16.5°C and climate was humid with rainforests because of the higher carbon dioxide levels. In the sea calcite precipitation was extensive.

Jurassic marine vertebrates mainly comprised fish with modern sharks and rays being common, and reptiles which included ichthyosaurs plesiosaurs, pliosaurs and crocodiles. Turtles were plentiful in lakes and rivers. Marine invertebrates included pelecypods, ammonites, belemnites, echinoids, crinoids, brachiopods, gastropods, sponges, coralline algae and a significant radiation of planktonic organisms.

Dinosaurs were the dominant, terrestrial animal life form in the Jurassic. Sauropods were herbivorous quadrupeds with very long necks, long tails, small heads and thick legs. They included Camarasaurus, Apatosaurus, Diplodocus and Brachiosaurus which was as huge dinosaur. They were preyed upon by large, meat eating theropods such as Ceratosaurus, Megalosaurus, Torvosaurus and Allosaurus, which was similar to the later evolving Tyrannosaurus rex. Other well known dinosaurs were the plated Stegosaurus and flying Pterosaurs. During the late Jurassic the first birds like

Archaeopteryx evolved from small coelurosaurian dinosaurs.

Early mammals were mainly small herbivores or insectivores and were not in competition with the larger reptiles. Placental mammals evolved in the mid-Jurassic.

Plant life comprised ferns and gingkoes. Gymnosperms which evolved included cone-bearing conifers which were pollinated by wind, cycads and Bennettitales. Smaller ferns were the dominant undergrowth and seed ferns were prevalent. Gingkoes were common in the mid to high northern latitudes.

The Jurassic-Cretaceous extinction was probably the result of intense volcanic activity caused by the break up of Pangaea and Gondwana. Massive extinction of sauropods and ichthyosaurs occurred, the latter being wiped out. The increase in sea levels resulted in the opening of the Atlantic Ocean which continued to get larger over time.

4.8.4 *Cretaceous*

The Cretaceous period which extended from 145 Ma to 65 Ma, had a relatively warm climate with high sea levels and numerous, shallow inland seas. The average surface temperature was 18°C. Mean atmospheric oxygen and carbon dioxide levels were 30% and 1700 ppm respectively. These levels were significantly higher than current levels.

Extensive chalk deposits formed in the Cretaceous. This rock type consists of coccolith skeletons of a type of algae. In many places black anoxic shales formed, being an important source rock for oil and gas.

The supercontinent Pangaea completed its break up in this period. Orogenies continued in North America as the Atlantic Ocean widened. Gondwana broke up into

4.0 THE EVOLUTION OF LIFE ON EARTH

South America, Antarctica and Australia rifted away from Africa, resulting in the formation of the South Atlantic and Indian Oceans. To the north of Africa the Tethys Sea continued to narrow. At the peak of the Cretaceous transgression about one third of the Earth's current land mass was under water.

In India massive basalt flows termed the Deccan Traps were extruded during the late Cretaceous and early Paleocene.

During this period flowering plants (angiosperms) spread, assisted by the coevolution of insects. Conjunctively the earlier Mesozoic gymnosperms continued to flourish. On land mammals were small and insignificant. The predominant fauna was archosaurian reptiles especially dinosaurs such as Tyrannosaurus rex, Carnotaurus, Triceratops and Tarbosaurus. Pterosaurs were common in the early and middle Cretaceous.

Fauna which became common in the seas included rays, modern sharks and teleost fish. Marine reptiles included ichthyosaurs, plesiosaurs and mosasaurs. Ammonites, clams, foraminifera and echinoderms flourished. Flightless, marine diving birds evolved.

At the end of the Cretaceous the Deccan Traps and other volcanic eruptions were poisoning the atmosphere. At this time a large asteroid about 10 km in diameter crashed into the Earth at the Yucatán Peninsula in southeastern Mexico, forming the 180 km wide Chicxulub Crater. This caused the Cretaceous-Paleogene (Tertiary) extinction event which resulted in the mass extinction of 75% of plant and animal species on Earth, including all non-avian dinosaurs. Other victims included mammals, pterosaurs, birds, lizards and plants. Marine life which was devastated included giant lizards, plesiosaurs, fish, sharks, molluscs and many species of plankton.

AN INTRODUCTION TO THE NEW WORLD ORDER

The Cretaceous-Paleogene extinction event is believed to be the result of three major causes which were volcanism, marine regression and extraterrestrial impact. The Deccan Traps volcanism included the release of dust and sulphuric aerosols into the atmosphere which blocked sunlight and reduced photosynthesis in plants. Increased carbon dioxide emissions would have increased the greenhouse effect once the dust and aerosols cleared.

In the final stage of the Cretaceous the sea level fell by more than any other time in the Mesozoic era. The cause is believed to have been mid-ocean ridges becoming less active and sinking. The severe regression greatly reduced the continental shelf area and resulted in the loss of epeiric seas such as the Western Interior Seaway of North America.

The effect of the asteroid impact was to inhibit photosynthesis by forming a dust cloud which blocked out sunlight for over a year and injecting sulphuric acid aerosols into the stratosphere which reduced the amount of sunlight reaching the Earth's surface by as much as 20%. It would take years for such aerosols to dissipate. Intense pulses of infrared radiation would have caused widespread fires. The high, atmospheric oxygen levels in the late Cretaceous would have facilitated combustion. The oxygen level plummeted in the early Cenozoic era. Widespread fires would have increased atmospheric carbon dioxide content which would have a temporary greenhouse effect.

The Cretaceous-Paleogene extinction had a significant effect on evolution with dinosaurs being replaced by mammals. Other groups such as snakes, lizards, fish and insects underwent major radiations.

4.0 THE EVOLUTION OF LIFE ON EARTH

4.9 Cenozoic

4.9.1 Introduction

The Cenozoic era extends from 66 Ma until the present and is divided into the Paleogene, Neogene and Quaternary periods. The Paleogene period which began 66 Ma and ended 23.03 Ma included the Paleocene, Eocene and Oligocene epochs. The Neogene period extended from 23.03 Ma to 2.588 Ma and comprises the Miocene and Pliocene epochs. The Quaternary period extends from 2.588 Ma to the present and is divided into the Pleistocene and Holocene epochs. The Cenozoic geological time scale with distinctive features is described in Table 10.

The Cenozoic is known as the Age of Mammals because the extinction of many groups including dinosaurs in the Cretaceous – Paleogene extinction event allowed mammals to greatly diversify.

The Earth's climate had begun a drying and cooling trend culminating in the Pleistocene glaciations. The continents moved into their current positions.

4.9.2 Paleocene

The Paleocene epoch extended from 66 to 56 million years ago. It began with the Cretaceous – Paleogene extinction event which wiped out non-avian dinosaurs, giant marine reptiles and much other fauna and flora. It ended with Paleocene – Eocene Thermal Maximum.

The early Paleocene climate was cooler and drier than that of the preceding Cretaceous period and temperatures rose markedly at the Paleocene – Eocene Thermal Maximum. The worldwide climate became warm and humid with extensive tropical forests and hot, arid equatorial areas.

AN INTRODUCTION TO THE NEW WORLD ORDER

Table 10. Cenozoic Geological Time Scale

Period	Epoch	Distinctive Features	Approx. Age (Ma)
Quaternary	Holocene	Quaternary Ice Age ends and interglacial begins, rise of human civilisation with agriculture and Stone, Bronze and Iron Ages, Sahara Desert forms, worldwide expansion of the human species, rise in atmospheric carbon dioxide from 280 ppm to 400 ppm following the Industrial Revolution	0.0000
	Pleistocene	Evolution and extinction of many large mammals, appearance of anatomically modern Homo sapiens, repeated glacial cycles and interglacials, start of human Stone Age culture with increasing technical complexity, eruption of Campi Flegrei supervolcano in Italy about 39 000 years ago resulting in mass extintion of Neanderthals, eruption of Lake Toba supervolcano in Sumatra, Indonesia about 75 000 years ago resulting in decrease in human numbers to about near extinction 3000 to 10 000 surviving individuals, Quaternary Ice Age continues, atmospheric carbon dioxide levels ranging from 200 to 300 ppm, Pleistocene ended with the Older Dryas and Younger Dryas cold periods	— 0.0117
Neogene	Pliocene	Beginning of Quaternary Ice Age, cool and dry climate, evolution of many existing mammals and recent molluscs, spread of Australopithecus, imminent appearance of Homo habilis, supernovae around two million years ago damaged Earth's ozone layer	— 2.588
	Miocene	Series of ice ages, evolution of modern mammals and birds, apes became diversified, ancestors of humans split away from the ancestors of chimpanzees, horses and mastodons (mammoths) flourished, grasses became prominent, whales, seals and kelp spread, widespread forests lowered atmospheric carbon dioxide levels from 650 ppm to 100 ppm, New Zealand, European, Carpathian and Hellenic orogenies, Middle Miocene disruption resulted in significant cooling and widespread extinctions	— 5.333
Paleogene	Oligocene	Cool climate trending towards icehouse, continents continued to drift to their present positions, rapid evolution and diversification of fauna, particularly mammals, expansion of angiosperms throughout the World, the La Garita volcanism in southwestern Colorado was the largest Cenozoic era eruption with the volcanic tuff volume being 5000 km^3	— 23.03
	Eocene	Wide range of climatic conditions ranging from the warmest temperatures of the Cenozoic era and ending in an icehouse climate, Eocene - Oligocene extinction event known as the Grande Coupure resulting in large scale extinction of aquatic organisms. Archaic mammals flourish, appearance of some modern mammal families, primitive whales diversify, first grasses, reglaciation of Antarctica, major settlement and decay of sea floor algae absorbing large quantities of atmospheric carbon dioxide, reducing the level from 3800 ppm to 650 ppm, Rocky Mountains orogenies in North America, the Alps orogeny in Europe, and Hellenic orogeny in Greece, Europe, Greenland and North America drift apart, India collides with Asia starting the folding of the Himalayas	— 33.9
	Paleocene	The Paleocene epoch started with the Cretaceous - Paleogene extinction event and ended with the Paleocene - Eocene Thermal Maximum where a significant rise in the atmosphere temperature resulted in the mass extinction of benthic foraminifera. Climate was mainly tropical, modern plants evolved, mammals diversified into a number of primitive groups following the extinction of the dinosaurs, first large mammals appeared, Alpine orogeny in Europe, warm seas circulated throughout the world	— 56.0 66.0

4.0 THE EVOLUTION OF LIFE ON EARTH

Cretaceous period plate tectonics continued into the Paleocene epoch. The supercontinent Lacrasia had not yet split into three continents and Europe and Greenland were still connected. A landbridge intermittently joined America and Asia while Greenland and North America commenced to rift. The Rocky Mountains orogeny continued in the American west and South and North America were separated by equatorial seas. Gondwanaland continued to split apart with Africa, South America, Australia and Antarctica separating. Africa headed north towards Europe, and India commenced to drift towards Asia ultimately forming the Himalayas.

Throughout the World there were circulating warm seas which promoted marine life including coral reefs. Sharks became the top predators while ammonites and numerous foraminifera species became extinct.

Modern plant species, cacti and palm trees appeared as did tropical, sub-tropical and deciduous forests. The extent of angiosperms continued to expand as did the co-evolution of pollinating insects.

Mammals evolved in the Triassic period and were small and generally nocturnal. These creatures managed to survive the Cretaceous – Paleogene extinction event. In the Paleocene they became bigger, more ecologically diverse and included monotremes, marsupials and placentals.

Reptiles were more prolific in the Paleocene because of the climatic conditions. Species included champsosaurs (aquatic reptiles which became extinct in the Miocene), crocodilia, soft shelled turtles, snakes and lizards.

A resurgence of birds occurred during the epoch. Large flightless and carnivorous birds evolved as well as types of owls.

AN INTRODUCTION TO THE NEW WORLD ORDER

The Paleocene – Eocene Thermal Maximum (PETM) occurred about 56 million years ago and lasted about 170 000 years. This event resulted in an initial 5°C temperature rise and occurred because of significant changes in the Earth's carbon cycle. A mass extinction of benthic foraminifera occurred. Acidification of deep waters which spread from the North Atlantic Ocean explains the spatial variations in carbonate solution.

At the beginning of the PETM the average, global temperature increased by 6°C in about 20 000 years. There was a rise from about 17°C to 23°C. Evidence suggests a carbon addition range from about 2000 to 7000 gigatons to accommodate the temperature increase.

During the PETM the climate became much wetter and ocean freshwater increased in the northern hemisphere. Many ocean basins remained bioturbated except in the North Atlantic Ocean where bottom water anoxia occurred. Extensive acidification of oceans occurred.

The extinction of 35% to 50% of benthic foraminifera was probably the result of reduction in oxygen availability and oceanic acidification. Locally deep-sea anoxia contributed to adverse conditions.

There was an increase in mammalian abundance with horses and primates spreading around the globe. Increased carbon dioxide levels may have promoted dwarfing which could have encouraged speciation.

The PETM was caused by a massive injection of carbon into the atmosphere. The source of the carbon has not been located. A possible source is methane clathrate (methane hydrate) which is produced by decomposing microbes in sea bottom sediments. As sea water temperature rises biogenic clathrates can release methane gas which finds its way into the atmosphere, facilitating temperature increase.

4.9.3 Eocene

The Eocene epoch continued from 56 to 33.9 million years ago. There was a wide range of climatic conditions ranging from the warmest temperatures of the Cenozoic era and ending in an icehouse climate. The PETM commenced 56 million years ago and increased during the Early Eocene Climatic Optimum (EECO) at about 51.5 to 50.9 million years ago, when there was no ice on Earth and similar temperatures at the equator and the poles. Temperatures then reduced to an icehouse climate at the Eocene – Oligocene transition, about 34 million years ago.

The greenhouse gases carbon dioxide and methane had a significant effect on Eocene temperature changes. During peak temperatures the maximum estimated atmospheric, carbon dioxide concentrations vary from 900 ppm to over 2000 ppm. Current carbon dioxide level is at about 400 ppm. The amount of oxygen in the atmosphere doubled at the beginning of the Eocene epoch.

During the early Eocene a significant amount of methane was released into the atmosphere from wetlands, swamps and forests. The amount of methane produced was triple the volume of current production.

At the end of the EECO the carbon dioxide level began to reduce due to increased production of siliceous plankton and marine carbon burial. The Azolla event occurred in the middle Eocene about 49 million years ago, when the freshwater fern Azolla proliferated in the Arctic Ocean. The modern fern Azolla filiculoides is a related species. The Azolla event occurred over about 800 000 years and coincides with a dramatic decline in carbon dioxide levels from a recorded maximum of about 4000 ppm to about 600 ppm. Azolla has unique properties and can draw down 0.4 tonne of nitrogen per hectare per year and 2.4 tonnes of carbon per hectare per year.

AN INTRODUCTION TO THE NEW WORLD ORDER

The Azolla bloom covered an area of 4 million km² of the Arctic basin. This event alone accounted for more than 80% decrease in carbon dioxide changing the climate from greenhouse to icehouse. The Arctic cooled from 13°C to today's -9°C. At the Eocene – Oligocene transition the atmospheric carbon dioxide concentration had reduced to about 700 ppm.

Early Eocene increases in polar temperatures can be attributed to increased carbon dioxide levels, ocean heat transport, oceanic hydrothermals and the presence of polar stratospheric clouds.

The Eocene was characterised by the warmest period in the Cenozoic and ended with an icehouse climate including rapid expansion of the Antarctic ice sheet. The cooling climate began about 49 million years ago and was largely due to the Azolla event, which resulted in major reduction of atmospheric carbon dioxide and temperature decrease. Global cooling continued until there was a reversal to warming around 42 million years ago. This event, known as the Middle Eocene Climatic Optimum was due to increases in atmospheric carbon dioxide. The warming was short lived and cooling returned about 40 million years ago. The cooling continued to the Eocene – Oligocene transition (33.9 Ma), when there was a massive expansion of the Antarctic ice sheet. A major extinction event called the Grande Coupure occurred at this time which mainly affected aquatic organisms. There was also a major European turnover in mammalian fauna and all European primates became extinct.

The continents continued to drift to their present positions in the Eocene. Australia and Antarctic parted around 45 Ma. The northern supercontinent of Laurasia began to break up into Europe, Greenland and North America. At about 35 Ma, an asteroid impact on the eastern coast of North America formed the Chesapeake

Bay impact crater. India began its collision with Asia to form the Himalayas.

At the beginning of the Eocene the Earth was entirely covered in forests. Cooling began about mid-period and by the end of the Eocene continental interiors had begun to dry out. Deciduous trees became better adapted to the environment and flourished over large parts of the northern continents. By the beginning of the Oligocene Antarctica hosted deciduous forests and large areas of tundra.

Modern mammals such as artiodactyls, perissodactyls and primates appeared in the early Eocene. Modern hoofed animals became prevalent in Europe and North America. Early forms of many other modern mammals appeared. Reptiles such as pythons and turtles were abundant as were insect faunas.

The Eocene oceans were warm and were teeming with fish, sharks and other fauna. Early whales evolved at this time.

The end of the Eocene was marked by the Eocene – Oligocene extinction event or the Grande Coupure. Large scale extinction of fauna and flora occurred during major climate cooling.

4.9.4 Oligocene

The Oligocene epoch extended from 33.9 Ma to 23 Ma. The Oligocene is considered to be the link between the old World of the tropical Eocene and the more modern ecosystems of the Miocene. Major Oligocene changes included global expansion of grasslands and contraction of tropical forests to the equatorial belt.

The Oligocene commenced with the Eocene – Oligocene extinction event as known as the Grande Coupure, in which Asian fauna replaced European fauna. The global climate change was minor with an abrupt change occurring between 28 and 26 Ma when the massive La Garita volcanism formed the La Garita Caldera in the San Juan Mountains in southwestern Colorado, United States. The La Garita Caldera is the largest known explosive eruption in Earth's history. The resulting Fish Canyon Tuff deposit has a volume of about 5000 km^3. The caldera is 35 by 75 kilometres in size and is an extinct supervolcano.

During this epoch the continents continued to drift to their current positions. Orogenies continued in western North America and the Alps in Europe. The Tethys Sea became isolated remnants as the African plate pushed north into the Eurasian plate. A land bridge existed between North America and Europe, South America detached from Antarctica allowing the Antarctic Circumpolar Current to flow, rapidly cooling this area.

Angiosperms continued to flourish and temperate deciduous forests expanded. Open plains and deserts became more common, as did grasslands.

Open landscapes allowed animals to grow to larger sizes. Numerous groups such as horses, entelodonts, rhinoceroses, oreodonts and camels adapted to the expanding plains. Cats which originated in Asia spread to Europe. There is a large gap in the Oligocene primate fossil record. Monkeys evolved during the early Oligocene. The marine animals of the Oligocene were similar to today's fauna.

The beginnings of modern ocean circulation commenced in the Oligocene. Drake Passage located between South America and Antarctica opened as did the Tasmanian Gateway between Australia and Antarctica. The collisions of the African plate with the European plate and of the Indian subcontinent with the Asian plate cut off the Tethys seaway.

4.0 THE EVOLUTION OF LIFE ON EARTH

Ocean-wide cooling occurred during the Oligocene. The main source of deep water was the North Pacific and the Southern Ocean. Deep water also commenced to form in the North Atlantic.

The Haughton impact crater is located on Devon Island in Northern Canada. It is about 23 km in diameter and formed about 23 million years ago by an impacting object about 2 km in diameter.

4.9.5 Miocene

The Miocene is the first geological epoch of the Neogene period and occurred from 23.03 Ma to 5.33 Ma. From the Oligocene to the Pliocene the Earth gradually cooled, but remained moderately warm.

The apes evolved and diversified in the Miocene. By the end of this epoch human ancestors had split away from the ancestors of the chimpanzees and continued to evolve. Grasslands continued to expand as in the Oligocene. Kelp forests flourished in the oceans. Plants and animals were modern with mammals and birds being well established.

The continents continued to drift to their current locations. The land bridge between North and South America had not formed. Orogenies occurred in western North America, Europe and East Asia. The Tethys Seaway disappeared as Africa collided with Eurasia in the Turkish–Arabian region. Temporary drying up of the Mediterranean Sea occurred at the end of the Miocene due to uplift in the western Mediterranean region and a global fall in sea level. This event is known as the Messinian Salinity Crisis (MSC). The Atlantic waters found a way through the Gibraltar Strait and refilled the Mediterranean 5.33 million years ago in an event known as the Zanclean flood.

Throughout the Miocene the climate remained moderately warm with slow global cooling tending to the Pleistocene glaciations. Miocene warming occurred between 21 and 14 million years ago and then there was a sharp drop in global temperature known as the Middle Miocene Climate Transition (MMCT). At about 8 million years ago there was another sharp drop in temperature resulting in expansion of the Antarctic ice sheet and the formation of large glaciers in Greenland.

The Miocene epoch was characterised by extensive grasslands and kelp forests. The presence of grasslands promoted the evolution of ungulate mammals. Marine and continental fauna were modern.

Early Miocene groups include entelodonts, nimravids, three-toed horses and oreodonts which became extinct in the early Pliocene. Miocene mammals included dogs, bears, raccoons, horses, beavers, deer, camels and whales as well as now extinct groups like borophagine dogs, gomphotheres, three-toed horses and hornless rhinos. About 100 species of apes ranged throughout Africa, Asia and Europe. The apes that gave rise to modern hominids, which are a taxonomic family of primates, lived between 7 and 8 million years ago. The first hominids, which were bipedal apes of the human lineage appeared in Africa at the end of the Miocene.

Most modern bird families were present in the Miocene and included ducks, plovers, owls, cockatoos and crows. The diversification of marine birds was prolific.

Algal kelp forests flourished and led to new species of sea life including otters, fish and invertebrates. Cetaceans were prevalent in the Miocene as were large sharks. Crocodilians diversified, growing to gigantic sizes.

Decrease in temperature during the Middle Miocene at about 15 Ma (MMCT) resulted in the growth of the East

4.0 THE EVOLUTION OF LIFE ON EARTH

Antarctica ice cap and oceans cooled, partly due to the formation of the Antarctic Circumpolar Current. The MMCT resulted in widespread extinction of terrestrial and aquatic life forms. This event resulted in major expansion of the East Antarctic ice sheet and formation of cold, Antarctic deep water.

4.9.6   Pliocene

The Pliocene epoch extended from 5.33 million to 2.58 million years ago. It is the second and youngest epoch of the Neogene period. The upper boundary of the Pliocene was set at the start of the Pleistocene glaciations.

The average global temperature in the mid-Pliocene was about 2 to 3°C higher than today and the global sea level was 25 m higher. The Arctic ice cap formed and global cooling resulted in the retreat of forests at the expense of grasslands and savanna.

Continents continued to drift close to their present locations and South America became linked to North America by the Isthmus of Panama. The Isthmus caused warm equatorial ocean currents to be cut off and the Atlantic cooling cycle commenced with the cold Arctic and Antarctic waters dropping temperatures in the now separated Atlantic Ocean. Africa's collision with Europe formed the Mediterranean Sea and the remnants of the Tethys Ocean disappeared. A land bridge formed between Alaska and Asia.

The Pliocene change to a cooler and drier climate reduced the extent of tropical forests. Deciduous forests were prolific and coniferous forests and tundra covered much of the north, while grasslands extended on all continents. Tropical forests were confined to the Equator and deserts appeared in Asia and Africa.

Marine and continental faunas were generally modern. The more advanced primates continued to evolve with the australopithecines, the first creatures to be termed human, appearing early in the epoch. Included are all species in the related genera of Australopithecus and Paranthropus, also known as robust australopithecine. The australopithecine species which was bipedal, evolved into the Homo genus in Africa. The extinct species of Homo habilis appeared in Africa about 2.4 million years ago.

In North America large mastodons, gomphotheres, rodents and opossums flourished while ungulate, hoofed animals such as camels, deer and horses receded. Extinction of rhinos, three toed horses, oreodonts, protoceratids, chalicotheres, Borophagine dogs and Agriotherium occurred. Carnivores such as weasels, dogs and bears diversified. Ground sloths, large glyptodonts and armadillos migrated north with formation of the Panama Isthmus.

In Eurasia elephants, gomphotheres, stegodonts, tapirs, rhinos, cows, antelopes and rodents were successful. Hyraxes migrated across from Africa and camels crossed into Asia from North America. Hyenas and sabre-toothed cats appeared while primate and horse distribution declined.

African evolution was dominated by hoofed animals and primates. Australopithecines which were among the first hominins, appeared in the late Pliocene. They were followed by Homo habilis of the Homo species. Populations of elephants, cows, antelopes and rodents were successful. Early giraffes appeared as did horses and modern rhinos. Bears, dogs and weasels joined cats, hyenas and civets as predators. Pigs continued to prevail.

North American fauna species which invaded South America included rodents, primates, weasels and bears. In South America litopterns and notoungulates became extinct but macrauchenids and toxodonts survived.

South American species which migrated north and thrived included glyptodonts, giant sloths, rodents, pampatheres and armadillos.

In Australia marsupials including kangaroos, wombats and the huge diprotodons were dominant. Carnivorous marsupials included dasyurids, thylacines and Thylacoleo. The first rodents arrived in Australia and the modern platypus appeared.

The remains of birds have been rarely encountered in rock strata. However birds were common and included modern species as well as some which have become extinct.

As the climate cooled, alligators and crocodiles died out in Europe. With the presence of more rodents and birds, venomous snake numbers increased. The modern alligator continued to thrive within a lesser range, as did giant tortoises in North America.

The Pliocene seas had abundant sea cows, sea lions and seals. Invertebrates continued to proliferate in particular molluscs which included gastropods and pelecypods.

About once every 50 years a massive star explodes in the Milky Way resulting in a huge release of energy known as a supernova. Around the end of the Pliocene a group of bright stars called the Scorpius-Centaurus OB association passed within 130 light years of the Earth with one or more supernova explosions forming a cavity known as the Local Bubble in the interstellar medium. The blast damaged the Earth's ozone layer and caused the extinction of some marine life.

4.9.7 Pleistocene

The Pleistocene epoch lasted from 2.588 million to 11 700 years ago and included repeated glacial cycles. The Pleistocene is the first epoch of Quaternary period and the sixth epoch of the Cenozoic era. It ended with the end of the last glaciation. The Pleistocene corresponds to the Paleolithic period which was the early phase of the Stone Age (the Old Stone Age) and extended from 2.5 million to about 10 000 years ago.

The modern continents were essentially in their current locations and since the Pleistocene to the Present interval, have not moved more than 100 km relative to each other. Oceanic currents have remained the same. The atmospheric carbon dioxide levels ranged from 200 to 300 ppm.

Repeated glacial cycles occurred in the Pleistocene with up to 30% of the Earth's surface being covered with ice. During glacial events the global sea level fell by a maximum of 100 m. Glaciers were prevalent in Europe, North America, Patagonia, parts of east and central Africa, New Zealand and Tasmania.

At least 11 major Pleistocene glacial events occurred as well as numerous minor events. Glaciation occurred as a series of glacials and interglacials caused by changes in climate. The main contributing factor is considered to be Milankovitch cycles. These are long term variations in the Earth's orbit which result in climate changes over periods of hundreds of thousands of years and cause ice ages.

The glacial periods of the Quaternary are named from the most recent to the oldest (Table 11). Dates are shown in thousands of years before the present (ka).

Marine and terrestrial faunas were modern and many mammals were much larger than their modern relatives. The severe climate changes caused by the ice age severely affected the fauna and flora. With each glacial advance continents became barren with plants and animals retreating towards the Equator. Severe climatic stress caused a reduction in living space and reduced

4.0 THE EVOLUTION OF LIFE ON EARTH

food supply. Mass extinction of large mammals occurred and included mammoths, mastodons, sabre-toothed cats, glyptodons, ground sloths, Irish elk and cave bears. Neanderthals also became extinct as did native horses and camels in North America. At the end of the last ice age cold blooded reptiles, smaller mammals like mice, migratory birds and swift animals like deer had replaced the megafauna and migrated north.

Table 11.   Chronology of Quaternary Glacial Cycles

Backwards Glacial Index	Name / Alpine	Interglacial/Glacial	Period (ka)	Epoch
		Interglacial	Present - 12	Holocene
1st	Würm	Glacial Period	12 - 71	Pleistocene
	Riss-Würm	Interglacial	115 - 130	Pleistocene
2nd	Riss	Glacial Period	130 - 200	Pleistocene
	Mindel-Riss	Interglacial (s)	374 - 424	Pleistocene
3rd - 6th	Mindel	Glacial Period (s)	424 - 478	Pleistocene
	Günz-Mindel	Interglacial (s)	478 - 563	Pleistocene
7th - 8th	Günz	Glacial Period	621 - 676	Pleistocene

Humans evolved into their present form during the Pleistocene. In the early Pleistocene Paranthropus and early human ancestors persisted but then became extinct. The only hominid species found in the Pleistocene fossil record is Homo erectus. In the middle and late Pleistocene types of humans appeared, together with more elaborate stone tools. Modern humans migrated out of Africa after the Riss glaciation in the middle Pleistocene and spread over the ice-free World during the late Pleistocene. Humans in this migration interbred with archaic human forms existing outside of Africa, incorporating archaic genetic material into the modern human gene pool.

The Toba supervolcano eruption occurred about 75 000 years ago at the present location of Lake Toba in Sumatra, Indonesia. It was one of the Earth's largest known eruptions, causing a global volcanic winter lasting six to 10 years and possibly a 1000 years cooling event. The eruptive volume was about 3000 km^3 of which about 800 km^3 was ash. The erupted mass was about 100 times greater than the largest volcanic eruption in recent history which was the 1815 eruption of Mount Tambora in Indonesia, causing the 1816 "Year Without a Summer". The Toba supereruption resulted in very extensive ash fall and the injection of vast quantities of noxious gases into the atmosphere. This eruption coincided with the onset of the last glacial period. Following this event the total human population decreased to about 3000 to 10 000. The cause was probably the destruction of food sources for humans.

The Older Dryas was a cold period which occurred about 14 000 years ago. The gradual global warming which occurred following the peak of the last glacial period about 22 000 years ago, was interrupted by three cold intervals. The Older Dryas was preceded by the Oldest Dryas and was followed by the Younger Dryas.

Older Dryas flora species included birch, pine, sagebrush and joint-fir. The Arctic plains were home to a number of animal species consisting of bison, deer, elk, musk ox, wild horse, woolly rhino, woolly mammoth, bear, hyena, cave lion, wolf and Arctic fox. Smaller animals which lived in the herbaceous blanket were the hare, lemming, vole, squirrel and jerboa. Maritime animals included the seal, walrus, beluga, killer whale and grey whale.

Eurasia was populated by Homo sapiens (modern man) known as the Cro-Magnon man in the late Pleistocene. The oldest dated remains are about 45 000 years old.

4.0 THE EVOLUTION OF LIFE ON EARTH

Cro-Magnons were tall, robustly built and powerful. An exodus from Africa over the Arabian Peninsula about 60 000 years ago brought modern humans to Eurasia.

Cro-Magnons were hunters killing mammoths, cave bears, horses and reindeer. They hunted with spears and javelins. Flax fibres were used for making cords, weaving baskets and sewing garments. They made flint tools and constructed shelters from rocks, clay, branches and animal hide.

In Europe Cro-Magnons co-existed with Neanderthals for over 10 000 years before the latter became extinct. Recent genetic studies indicate that some hybridisation with archaic humans occurred after modern humans emerged from Africa.

The extinction of Neanderthals occurred when the supervolcano Campi Flegrei erupted in Italy about 39 000 years ago. This volcanic eruption which was the most powerful in the last 200 000 years, changed the climate and spread volcanic ash throughout Europe. Most Neanderthals, except for a small group at Gibraltar, became extinct. The latter group also eventually became extinct.

The Younger Dryas cold period, known as the Big Freeze, occurred between 12 900 and 11 500 years ago. Two possible causes for this event include collapse of the North American ice sheets or a meteor collision which caused cooling by dust accumulation in the stratosphere.

During the Younger Dryas glacial conditions rapidly returned to the higher latitudes of the northern hemisphere. In the southern hemisphere and in western North America the effects were less intense than in Europe. A number of changes occurred. In Scandinavia forests were replaced with glacial tundra. Glaciation and increased snow occurred on mountain ranges around the World. In North America the Clovis (paleo-indian) culture declined and animal species became extinct. There is evidence that a solar flare may have caused megafaunal extinction. The Laacher See volcano in Germany erupted at this time ejecting 10 km^3 of material and causing significant temperature changes in the northern hemisphere.

The Younger Dryas occurred over 40 to 50 years with three distinct steps each lasting about five years. This event is thought to have brought about cereal cultivation.

In the Quaternary period extinctions of many megafaunal species occurred in the late Pleistocene and Holocene. Such events occurred in North and South America, Australia and Europe. The three main explanations for the extinctions are climate change associated with ice, extermination by humans and changes in vegetation from grasslands to birch forests.

Many large animals disappeared from Africa and Asia throughout the Pleistocene epoch. Australian extinctions coincided with the first arrival of humans about 50 000 years ago. Extinctions which were common in northern Eurasia included the woolly mammoth and Neanderthals who survived until about 28 000 years ago. During the last 60 000 years about 33 genera of large mammals have become extinct in North America. During the Pleistocene South America remained largely unglaciated but at the start of the Holocene all megafauna became extinct.

There are various hypotheses for the Quaternary extinction event. The hunting hypothesis suggests that humans hunted megaherbivores to extinction which resulted in the death of carnivores and scavengers from lack of prey. The climate change hypothesis surmises that the ice ages caused the disappearance of certain animals. The continentality theory refers to hotter summers and colder winters causing extinctions. Increased continentality would have changed vegetation and resulted in reduced and less predictable rainfall. The hyperdisease hypothesis explains the extinction of large mammals during the late Pleistocene to highly virulent diseases brought by newly arrived humans and their domesticated animals. The second order predation scenario refers to the killing of predators by humans, resulting in boom and bust cycles in herbivorous populations. This caused changes in vegetation. The comet hypothesis suggests that a swarm of comets which occurred about 12 900 years ago, caused mass extinction. Evidence for such an event is not strong. There was probably a number of causes for the Quaternary extinction event, which resulted in the disappearance of a number of animal species.

4.9.8   Holocene

The Holocene is the most recent geological epoch of the Quaternary period and extends from about 11 700 years ago to the present time. It appears to be an interglacial phase of the current ice age.

The Holocene is characterised by the enormous growth and impact of the human species worldwide. Dramatic human expansion has affected the environment, other living species and the atmosphere.

There are no defined faunal stages for the Holocene. A subdivision which has been compiled based on human technological advancement includes the Mesolithic and Neolithic Stone Ages, Bronze Age and Iron Age. The time periods for these ages vary for different locations around the World.

Plate tectonic, continental movements during the Holocene have been less than one kilometre. Melting polar ice caused the sea level to rise about 35 m in the early Holocene. In the northern hemisphere post-glacial isostatic readjustment has resulted in land rising up to 180 m and in places the land is still moving.

Rising sea level resulted in temporary marine incursions into North America. Post-glacial rebound in Scandinavia resulted in the formation of the Baltic Lake which connected to the sea about 10 000 years ago to form the Baltic Sea. Rebound of Hudson Bay in North America caused it to reduce in size to its current boundaries.

During the Holocene the climate has been warm and stable. The current warming is an interglacial period and may not be the end of the current ice age.

At present the carbon dioxide level in the Earth's atmosphere is about 400 ppm. Since the industrial revolution the level has increased from 280 ppm to 400 ppm which is causing global warming. Over the past 400 000 years the atmospheric carbon dioxide concentrations have varied from about 180 ppm during extensive glaciations to 280 ppm during interglacial periods. Carbon dioxide levels reached a concentration of about 6000 ppm in the Cambrian and 7000 ppm prior to the Ordovician-Silurian extinction events. Past increases in carbon dioxide levels were due to significant volcanism. The current increase is due to human intervention.

Increases in atmospheric carbon dioxide levels lead to the greenhouse effect which causes rises in global temperatures. The results are higher rainfall and accelerated melting of the polar ice caps. Current carbon dioxide increases in the atmosphere are mainly caused by human industrial emissions. Any major volcanic activity would accelerate the rising atmospheric, carbon dioxide level. If the polar ice caps completely melt the global sea level would rise by about 70 m. Melted ice from Antarctica would result in a 61 m rise and the melting of Greenland and North Pole ice would cause a 9 m rise. Many coastal low land areas around the World would be flooded.

4.0 THE EVOLUTION OF LIFE ON EARTH

There has been no noticeable animal or plant evolution in the Holocene. In the late Pleistocene and early Holocene a number of large animals including mammoths, mastodons, sabre-toothed cats and giant sloths became extinct. In North America the disappearance of horses and camels coincided with the arrival of ancestors of the Amerindians and possibly a bolide impact which could have contributed to the Younger Dryas.

The Holocene is characterised by human dominance. The beginning of this epoch corresponded to the Mesolithic Stone Age which was followed by the pre-pottery and pottery Neolithic Stone Age. More recent technological advancements resulted in the Bronze and Iron Ages. The late Holocene brought hunting improvements such as the bow and arrow.

The first humans were hunters and gatherers who survived on natural vegetation and game. They did not have permanent settlements and migrated as the seasons and climate changed. As the climate became warmer between 10 000 and 7000 years ago, people adapted by domesticating plants and animals. Agriculture became the main human activity for food production and communal groups became more sedentary forming permanent settlements. This is known as the Neolithic Revolution or the Agricultural Revolution.

Throughout the history of the Earth there have been seven major extinctions (see Section 4.11) and the Holocene extinction is sometimes called the sixth extinction of the Phanerozoic eon. This event is mainly due to human activity. The present rate of extinction may be up to 140 000 species each year. It has been estimated that by the end of the 21st century up to 50% of all living species of flora and fauna on Earth will be extinct. Extinctions include plants, mammals, birds, amphibians, reptiles and arthropods. Examples are the woolly mammoth, American mastodon, moa, aurochs, dodo, quagga, thylacine, Caribbean monk seal and Yangtze River dolphin. About 7% of all species on Earth have already been lost.

Current extinctions are due to an exploding human population which is rapidly changing the planet's environment. This type of extinction which commenced about 12 000 years ago, is now exponentially increasing.

4.10   Human Evolution

Human evolution is the process whereby genetic variation has resulted in the appearance of anatomically modern Homo sapiens. According to genetic studies, primates diverged from other mammals about 85 million years ago in the Late Cretaceous period. The earliest fossils occur in the Paleocene about 56 million years ago. Primates are characterised by having larger cerebral hemispheres than other mammals as well as an advanced development of binocular vision at the expense of smell, and specialisation of the appendages for grasping. They include apes, monkeys and related forms such as lemurs and tarsiers.

The order Primates is divided into two main groups which include prosimians and anthropoids (simians). Prosimians have characteristics more like those of the earliest primates and include lemurs, lorisoids and tarsiers. Simians include apes, monkeys and hominins, which are a group consisting of modern humans, extinct human species and all human immediate ancestors.

Human evolution is characterised by significant anatomical changes which have occurred following the separation of the last common ancestor of humans and the chimpanzees. Bipedalism is a basic hominin adaptation which has resulted in a number of skeletal changes to the legs, pelvis, vertebral column, feet, ankles and skull. Encephalisation is the development of a much larger brain in the human species than in other primates,

4.0 THE EVOLUTION OF LIFE ON EARTH

resulting in increased intelligence. The reduced degree of sexual dimorphism is visible in humans. Evolution of humans has resulted in increased importance of vision rather than smell, a smaller gut, loss of body hair, development of sweat glands and a chin, and formation of a descended larynx.

According to archeological evidence, the human species evolved from hominoids which appeared about 20 million years ago. Hominoids belong to the superfamily Hominoidae, which includes both humans and the apes and their extinct evolutionary precursors. The ancestors of modern humans branched off from the ape family between 5 and 8 million years ago. They are known as hominids and belong to the family Hominidae, which includes both extinct and modern humans as well as the Great Apes.

Australopithecines were an evolutionary stage of extinct hominids which existed in Africa between 1 and 4 million years ago. They were characterised by small cranial capacity, protruding jaws and upright stance. Australopithecines came in two forms which were small and lightly built (gracile) and more ruggedly built (robust). Australopithecus afarensis (occurring from 2.5 million to 4 million years ago) and Australopithecus africanus (occurring from 2.5 million to 3 million years ago) are classified as gracile, while Paranthropus (or Australopithecus) bolsei (occurring from 1.2 million to 2.6 million years ago) and Paranthropus (or Australopithecus) robustus (occurring from 1 to 2 million years ago) are described as robust.

The genus Homo appeared 2 million years ago, possibly as an offshoot from Australopithecines. The two forms appear to have co-existed for about 1 million years. Human culture developed as early hominids began making stone tools. Earliest fossil stone tools date back 2.5 million years.

AN INTRODUCTION TO THE NEW WORLD ORDER

A new type of hominid, known as Homo habilis, appeared in the fossil record of eastern and southern Africa with a date of 2 million years ago. It was considered to be intermediate between Australopithecines and Homo sapiens and persisted to about 1.6 million years ago.

Approximately 1.8 million years ago, a new form of early human appeared in the eastern African fossil record. This species is called Homo erectus and as well as surviving in Africa for over a million years, Homo erectus had also spread to Asia and Southern Europe by about 1 million years ago. This species died out about 300 000 years ago.

At about 400 000 years ago significant evolutionary changes had occurred in human populations to warrant recognition of a new species, known as archaic Homo sapiens. These changes mainly related to skull and braincase size. The features were intermediate between Homo erectus and modern Homo sapiens. This species persisted in Africa, Europe and Asia to about 100 000 years ago.

Neanderthals were an evolutionary line of humans which lived in Europe and Western Asia between 150 000 and 30 000 years ago. Their features included stocky build, large heads, large brains and large projecting noses. They are generally considered as a distinct species, known as Homo neanderthalensis. The extinction of Neanderthals was probably due to climate change and competition caused by the appearance of modern Homo sapiens.

Modern Homo sapiens appeared in Africa between 200 000 and 150 000 years ago. This species had a modern skeleton and progressed out over Africa, to Europe and Asia. Modern Homo sapiens had spread to the Middle East by 100 000 years ago; to East Asia by 70 000 years ago; to Australia by 50 000 years ago; to Europe by 40 000 years ago; to Siberia by 25 000 years ago; and to the

4.0 THE EVOLUTION OF LIFE ON EARTH

Americas by 15 000 years ago. Populations progressively migrated southwards in the Americas to present day Chile and Southern Argentina.

The terms Paleolithic, Mesolithic and Neolithic define prehistoric periods of human evolution. There is some variation of these periods throughout the World. The Paleolithic or Old Stone Age extended from about 2.5 million years ago to 10 000 years ago. The Mesolithic or Middle Stone Age ranged from 10 000 to 5000 years ago. The Neolithic or New Stone Age occurred from about 7000 to 3000 years ago and overlapped the Bronze Age which commenced 4500 years ago and the Iron Age which began about 3200 years ago.

By about 40 000 years ago which was the beginning of the Upper Paleolithic, Homo sapiens closely resembled modern humans. These early moderns who are known as Upper Paleolithic peoples, include the European Cro-Magnons and are termed European early modern humans. They lived from about 40 000 to 10 000 years ago. Cro-Magnons were strongly built and powerful with bodies that were heavy and solid, having a well defined musculature. The forehead was straight rather than sloping like in the Neanderthals and the brow ridge was slight. The face was short and wide with a prominent chin. Their vocal chords were like those of present day humans and they could probably speak. The brain capacity was larger than the average for modern humans, being about 1600 cubic centimetres. Cro-Magnons were anatomically modern, straight limbed and tall compared to the contemporaneous Neanderthals. Their average height was about 176 cm with large males being up to 195 cm tall. The earliest known Cro-Magnon remains have been discovered in Italy, Britain and the European Russian Arctic. Remains have also been found in France and Germany. Anatomically modern humans emerged from East Africa about 100 000 to 200 000 years ago. Around 60 000 years ago they spread from the Arabian Peninsula to Eurasia.

AN INTRODUCTION TO THE NEW WORLD ORDER

Like most early humans Cro-Magnons were nomadic or semi-nomadic and primarily big game hunters. They hunted with spears, javelins and spear-throwers.

In the Upper Paleolithic (Late Stone Age) which ranged from 40 000 to 10 000 years ago, there was a greater diversity of tools including the invention of the spear, bow and arrow. New techniques of toolmaking were introduced. Cultural advancement included the first known cave paintings. Human migration was active during this period with Australia and America being colonised. People lived in houses which were generally semi-subterranean huts with dugout floors, heaths and windbreaks.

The Mesolithic or Middle Stone Age commenced about 10 000 years ago, as the World's glacial conditions were moderating. The Paleolithic-Mesolithic boundary may vary by as much as several thousand years depending on the locality. The Mesolithic ended about 5000 years ago. Changes in climate required human populations to develop new hunting techniques including fishing and to collect wild plant foods. Domestication of plants and animals commenced. New tools and weapons were invented including ground stone axes and adzes, and microlithic blades for arrows, sickles, spears and daggers.

Advanced tool technology allowed Mesolithic populations to increase and migrate to diverse environments. Biological changes included a reduction of the human face to modern proportions and a reduction of body size in men resulting from new, less physical hunting techniques.

The Neolithic or New Stone Age began about 7000 years ago and ended about 3000 years ago when metal tools became widespread in the Bronze Age and the Iron Age. The Neolithic was characterised by the commencement of agriculture, pottery and sedentism. Two toolmaking

4.0 THE EVOLUTION OF LIFE ON EARTH

changes which occurred were the appearance of polished stone tools such as axes and a reduced reliance on prismatic blades. The Stone Age ended with the introduction of metal technology which commenced with copper working.

The Neolithic Revolution or Agricultural Revolution involved the large scale transition of many human cultures from a lifestyle of hunting and gathering to one of settlement and agriculture, which allowed increasingly large populations to be supported. These societies modified the natural environment by extending specific food crop cultivation through irrigation and deforestation.

Centres of early Neolithic agriculture included Central Africa, China, Central and South America, and Southwest and Southeast Asia. From these areas food production spread to most parts of the World. Southwest Asia was one of the first agrarian regions, with the remains of domesticated plants and animals, older than 7000 years, having been found in Israel, Jordon, Syria, Turkey, Iraq and Iran. The remains of small farming communities, which consisted of clusters of mud houses, have also been found in these areas.

As agricultural expansion gained momentum selection breeding of cereal grasses such as emmer, einkorn and barley occurred. Domestication of flax pea, chickpea, bitter vetch and lentil came later. Rye was introduced into Europe. Rice was successfully cultivated in Asia. Early crops also included figs, wild barley and wild oats. With animals, dogs were domesticated first followed by goats and sheep.

Early Neolithic harvesting tools were made of wood or bone with serrated stone blades. Implements include scythes, forks, hoes and ploughs. Mortars and pestles were used for grinding grain. Domesticated cattle were trained as draught animals for ploughing about 8500 years ago. Pottery was extensively used in the Neolithic for transporting and storing food, cooking, pipes and lamps. Original pottery was made by hand. Much clothing was made from animal skins but the introduction of woven textiles from flax, cotton and wool for clothing occurred in the Neolithic.

During most of the Neolithic age people lived in small tribes. There is no evidence of religion or government in the social structure of Neolithic society. Families and households were economically independent and the household was the centre of life.

In the Neolithic there was a significant change in shelter with the appearance of mud brick houses coated with plaster. Stilt house settlements were common in some areas.

The Bronze Age is characterised by the use of bronze and is the second period of the three-age Stone-Bronze-Iron system. It extended from about 4500 to about 2500 years ago. Tin and copper were mined and smelted to make bronze alloy.

The Iron Age followed the Bronze Age and generally ranged from about 3200 to 2000 years ago. Local variations occurred showing both the continuity and discontinuity with the previous late Bronze Age. In the Iron Age iron or steel in the most part replaced bronze in implements and weapons.

Civilisation which is defined as an advanced state of human society in which there are higher levels of human expression emerged as Neolithic villages expanded into towns and then cities. The first cities were in Mesopotamia (Iraq), followed by Egypt and the Indus Valley, between 6000 and 4500 years ago. Civilisation commenced in China about 4000 years ago, and in Central America and Peru between 3000 and 2000 years ago. The four areas known as cradles of civilisation were

4.0 THE EVOLUTION OF LIFE ON EARTH

in Mesopotamia, Egypt, the Indus Valley and the Yellow River Valley. These societies developed around rivers which provided enough water for large scale agriculture.

The transformation from village to city life resulted in agricultural innovation, diversification of labour to provide craftsmen and technicians, the development of centralised governments and the introduction of social stratification. Religion appears to have developed as means for coping with uncertainties in life.

Problems associated with early civilisations included waste disposal, pollution, disease, crowding, social inequalities and warfare. Infectious diseases have had a profound influence on human evolution and culture. The increased incidence of disease and mortality can be related to a new mode of life in Neolithic communities with sedentary life in fixed villages bringing sanitation and disease problems. Airborne diseases and the close association between humans and their domestic animals were conducive to the transmittal of some animal diseases to humans. Diseases acquired from domesticated animals included measles (cattle, rinderpest), tuberculosis (cattle), smallpox (cattle), influenza (pigs, ducks, chickens) and pertussis or whooping cough (pigs, dogs). A number of life threatening diseases such as smallpox, chicken pox and other childhood infectious diseases were not counteracted by medical science until the second half of the 20th century.

Human populations have developed with biological variety. In different parts of the World humans have adjusted to the physical environment in different ways. Such variation is the outcome of the evolution of modern Homo sapiens.

Modern humans can be classified as one of four groups. Caucasoids are from northern Europe to northern Africa and India. These are depigmented to a greater or lesser degree. Hair in males is generally well developed on the face and body, and is mostly fine and wavy or straight. A narrow face and prominent narrow nose are both typical. Negroids or Congoids are from sub-Saharan Africa. Skin pigment is dense and dark, hair woolly, noses broad, faces generally short, lips thick, and ears are squarish and lobeless. Stature varies greatly, from pygmy to very tall. The most divergent group are the Bushman and Hottentot peoples of southern Africa. Mongoloids are found in all of Asia except the west and south (India), in the northern and eastern Pacific, and in the Americas. The skin is brown to light, hair coarse, straight to wavy, and sparse on the face and body. The face is broad and tends to be flat. The eyelid is covered by an internal skinfold in the central populations but such folds are less marked or absent elsewhere. The teeth often have crowns more complex than in other peoples, and the inner surfaces of the upper incisors frequently have a shovel appearance. The Chinese, Koreans and Japanese are the typical populations. In central and northeastern Asia and among Inuit the flatness of face and nose is still more marked. In more marginal populations, such as the Ainu of Japan, aboriginal Taiwanese, Philippine Islanders, Indonesians and Southeast Asians these traits are less marked. The same is true of Polynesians and Micronesians. In American Indians, the face is usually broad but nasal bridges are apt to be more prominent relative to the eyes. The teeth are complex in pattern and shovelled incisors are prevalent. Australoids are the aboriginals of Australia and Melanesia. Their skin is dark and hair is predominantly wavy (Australia) or frizzy (Melanesia), with blondness in children (lost in adulthood) being common throughout. The head is long and narrow, the forehead sloping with prominent brow ridges, and the face has a projecting jaw.

Present human populations are primarily descended from one main population which originated recently in a single region and then migrated outwards. Current racial variations are predominantly a response to different environmental conditions which have resulted in physical changes in the human species.

4.11 Extinction Events and Implications for the Human Race

4.11.1 Introduction

An extinction event is an extensive and rapid decrease in the amount of life on Earth. It is identified by a significant change in the diversity and abundance of multicellular organisms. More than 99% of documented species are now extinct.

Seven documented, mass extinction events on Earth are described. Six of these events have occurred in the Phanerozoic eon. There were probably multiple mass extinctions in the Archean and Proterozoic eons, but before the Phanerozoic there were no animals with hard body parts to leave a significant fossil record.

The Great Oxygenation Event which occurred between 2.3 and 2.5 billion years ago, released a considerable amount of atmospheric oxygen. This resulted in the extinction of most anaerobic bacteria, which were eventually replaced by aerobic organisms. The Ordovician-Silurian extinction event that killed off 57% of all genera is ranked as the second largest of all extinctions. It occurred from 440 to 450 million years ago and was caused by a fall in atmospheric carbon dioxide level resulting in widespread glaciations. An alternative extinction possibility is that a gamma-ray burst destroyed the ozone layer exposing Earth's life to radiation and initiating global cooling. The Permian-Triassic extinction event which happened about 251 million years ago was the greatest such event recorded in paleontology, with 95% of all marine species and 70% of all terrestrial organisms becoming extinct. The main cause was catastrophic volcanic activity in Siberia and China which resulted in massive flood basalt eruptions. Vast quantities of carbon dioxide were released. Ocean venting of hydrogen sulphide may have also occurred and there are indications that a massive asteroid hit Antarctica at the end of the Palaeozoic era. Life on Earth recovered quickly after the Permian extinctions. The Triassic-Jurassic extinction event has been dated at 205 million years ago and resulted in the death of all archosaurs, synapsids, almost all amphibians and 34% of all marine life. The cause was probably massive volcanic eruptions resulting in the flood basalts of the Central Atlantic Magmatic Province. Ash and gases such as carbon dioxide and sulphur dioxide would have been released into the atmosphere. The Cretaceous-Paleogene extinction event which occurred 65 million years ago, is believed to be the result of three major causes which comprised volcanism, marine regression and extraterrestrial impact. There was mass extinction of 75% of plant and animal species on Earth which included all non-avian dinosaurs, mammals, pterosaurs, birds, lizards, plants, plesiosaurs, fish, sharks, molluscs and many species of plankton. At this time a large asteroid about 10 km in diameter crashed into the Earth at the Yucatán Peninsula in southeastern Mexico. The seventh mass extinction event has begun with the cause being the explosive human population increase which has now reached a total of 7.4 billion beings on Earth. Animals are becoming extinct at 100 to 1000 times faster than the normal background extinction rate. The present rate of extinction may be up to 140 000 species per year, which is faster than the Cretaceous-Paleogene extinction which wiped out the dinosaurs. The current mass extinction is the result of five main human activities. Human overpopulation is out of control. Most of the human population of 7.4 billion has no useful role, only indulging in rampant consumerism, lacking in productivity for the future of the human race and senseless breeding which is destroying the human gene

4.0 THE EVOLUTION OF LIFE ON EARTH

pool. A result of human overpopulation is habitat destruction and human-induced climate change which are accelerating the extinction of many species. Other related activities are widespread, planetary pollution and over-harvesting which includes hunting, fishing and gathering. The human race has also introduced invasive species which have displaced native species through predation, competition and disease.

A number of lesser extinctions have occurred throughout the Earth's history. These events which occur periodically every 26 to 30 million years, can be attributed to volcanic activity and climate change, bolide (meteor) impacts and ocean anoxia, which is depletion in the level of sea water oxygen. Changing atmospheric carbon dioxide levels have had a significant effect on climate.

4.11.2 Effects of Extinction Events on the Human Race

It is inevitable that the inner planets of the Solar System will be destroyed in about 5 billion years when the Sun will become a red giant engulfing the Earth. Prior to that event mass extinctions are possible on Earth from a number of other causes. The current human population on Earth is 7.4 billion, with an uncontrolled rate of increase. The planet is totally unprepared for any extinction event which could decimate the human race.

The extinction event with the highest probability of occurring is the eruption of a supervolcano which can affect climate over a significant area. There are 23 supervolcanoes around the World. The most powerful is in Yellowstone National Park, Wyoming, United States. This supervolcano has erupted three times in the last 2.1 million years with the last eruption occurring about 640 000 years ago. A future eruption could cover about half of the United States in ash, with disastrous consequences. The Campi Flegrei supervolcano is in southern Italy next to Naples. When it erupted 39 000 years ago, this was one of the most powerful supervolcano eruptions in the last 200 000 years. It resulted in a European volcanic winter which led to the extinction of the Neanderthals. The Tobu supervolcano eruption occurred about 75 000 years ago at the present location of Lake Tobu in Sumatra, Indonesia. It caused a global volcanic winter lasting six to 10 years and possibly a 1000 year cooling event. Following this event the total human population decreased to about 3000 to 10 000, brinking on extinction.

A major supervolcano eruption would decimate the human race. No contingency plan has been made to support a population of 7.4 billion with the advent of a volcanic winter accompanied by loss of food production. Mass, planetary starvation could result in extensive, human cannibalism. With continued uncontrolled population increase a human extinction event would become chronic. The solution to this problem lies in immediate birth control. There is no reason for the human population to exceed 2 billion on Earth. Even a population of this size would have difficulty coping with a volcanic winter. Birth control can be achieved by restricting all couples to one child. With the advance of genetic engineering and the development of human embryo farms the quantity and quality offspring can be properly regulated.

In preparation for a volcanic winter food storages are required. As well as durable stored foods, fungal spores would assist in survival through extended periods of unfavourable conditions. During a volcanic winter, photosynthesis is restricted due to insufficient sunlight. The alternative is chemosynthesis which uses energy released by inorganic chemical reactions to produce food. Chemosynthetic bacteria make organic matter by oxidising hydrogen sulphide or methane. Research and development of chemosynthetic food should be an urgent priority, initially on Earth and in future in extraterrestrial environments.

4.0 THE EVOLUTION OF LIFE ON EARTH

Meteor (bolide) collisions with Earth are the second most common cause of extinction events and they have occurred repetitively throughout Earth's history. The collision which produced the Cretaceous-Paleogene event and resulted in the extinction of the dinosaurs is the most recognised such event. The probability is high that future collisions will occur between meteors and comets and Earth, resulting in significant climate change. For protection of the human race an Earth shield is required to prevent large, extraterrestrial collisions. The shield should comprise the largest possible hydrogen bombs mounted on rockets. These would be fired at an incoming meteor causing fragmentation and reducing impact damage. The effectiveness of large hydrogen bombs can be assessed by shooting them at the Moon and measuring sizes of the craters formed by the detonations. Other methods for deflecting incoming meteors away from Earth also need to be investigated.

With the presence of such a large human population on Earth the outbreak of a non-curable, contagious disease could have disastrous consequences. Emphasis needs to be placed on genetic engineering so that antidotes can be quickly developed to combat new diseases. Populations should be able to be easily segregated to prevent disease from rapidly spreading.

5.0 THE NEW WORLD ORDER

5.1 Introduction

The purpose of the New World Order is to ensure the indefinite, dignified survival of the human race in a physically hostile Universe. A huge effort is required for mankind to become master of his own destiny. Major advancements in science and technology are necessary to prevent human extinction. Society needs to be totally focussed on human survival.

Until now the human race has evolved through a process of natural selection without any genetic programming of the human gene pool. Because of uncontrolled breeding, the human population has reached 7.4 billion with many individuals having defective genes. Without genetic programming the human gene pool will continue to deteriorate.

A population of 7.4 billion is unsustainable and is already destroying the environment on Earth. Natural resources are being depleted at an accelerated rate. The population needs to urgently be reduced to no more than 2 billion people to maintain environmental equilibrium and survival of life on the planet.

Current trends in sociological advancement and infrastructure development are being bogged down in a morass of bureaucracy and incompetence. At present the emphasis is to attempt to provide infrastructure for a burgeoning, unsustainable population at the expense of scientific advancements for the future survival of the human race. Continuation of this folly will result in the destruction of Earth's resources without any future human gain. By ignoring the effects of past extinction events and by blindly stumbling along, the human race exposes itself to catastrophic destruction.

5.2 Human Population

The current human population is 7.4 billion which is estimated to increase to 11.2 billion by the year 2100, given present trends. The birth rate is now about 135 million per year. The two most populous countries are China and India with 1.38 billion and 1.32 billion of individuals, respectively.

The human population has grown slowly for most of its existence on Earth. Threats such as disease and climate fluctuations kept life expectancy short and death rates high in pre-industrial society. At the beginning of the Industrial Revolution life expectancies were low in Western Europe and the United States due to harsh conditions and poor nutrition. From about 1850 to 1950 improvements in health and safety significantly improved living conditions. Changes which contributed to greater life expectancies included improved urban sanitation and waste removal, provision of better quality water supply, prevention of the transmission of infectious diseases through the development of vaccines and antibiotics, adopting workplace safety laws and implementation of better nutrition. As shown in Table 12, population growth accelerated very rapidly after 1804.

The current global sex ratio is about 1.01 males to 1 female and the average, global life expectancy is 70.5 years. Since 1804 (Table 12), the population has increased exponentially and is expected to reach 11.2 billion by 2100.

Table 12.   World Population Trend

World Population	Year	Intervening Years
5 million	8000 BC*	
200 million	AD*1	8001
1 billion	AD 1804	1803
2 billion	AD 1927	123
3 billion	AD 1960	33
4 billion	AD 1974	14
5 billion	AD 1987	13
6 billion	AD 1999	12
7.4 billion	AD 2016	17

* BC Before Christ, AD After Christ

There are two main problems with the current population trend. The population of 7.4 billion is excessive and unsustainable. If a mass extinction event such as an exploding supervolcano or terrestrial impact with a large meteor occurs, the resultant climate change would cause mass starvation of the human race. As well, the human gene pool is being degraded by random interbreeding. The four groups of modern humans which are classified as Caucasoid, Negroid, Mongoloid and Australoid all have genetic strengths and weaknesses. Random interbreeding will not improve the human gene pool.

In the New World Order the maximum recommended human population on Earth is 2 billion people. Even this number would be difficult to maintain in a mass extinction event with the added advantages of large storages of food and the ability to produce chemosynthetic protein. The required population target

5.0 THE NEW WORLD ORDER

can initially be achieved by birth control where each couple is restricted to one offspring. In the future the human population will be controlled by total embryo production and development in embryo farms. The current human gene pool is the result of natural selection through random interbreeding and is flawed. In the New World Order it is essential that human evolution is accelerated for ongoing survival of the species. This can be done by genetic programming of the human genome. Prior to such a technique being successfully developed it would be advantageous for the DNA (deoxyribonucleic acid) of brilliant individuals to be cryogenically preserved in liquid nitrogen and then cloned in the future. Cloning and genetic programming will facilitate evolution of the human race to ensure that offspring are designed for specific roles in a rapidly expanding technological society and for adaptation for space travel.

5.3 Genetic Engineering and Cloning

Genetic engineering is the process of directly manipulating an organism's genome using biotechnology. The goal is to change the genetic makeup of cells. Cloning is the process of producing populations of genetically identical individuals. These techniques are essential to be able to genetically program and artificially accelerate the evolution of the human race. Indefinite human survival is totally dependent on rapid advances in science and technology to cope with changing physical conditions in the Universe.

DNA is the molecule that contains and transfers most of the genetic instructions used in the development, functioning and reproduction of all living organisms. The DNA molecule consists of two biopolymer strands (polynucleotides) coiled around each other to form a double helix. The strands are made up of nucleotides each of which is composed of a nitrogen containing nucleobase consisting of either adenine (A), cytosine (C), guanine (G) or thymine (T) as well as deoxyribose and phosphate. DNA stores biological information which is replicated as the two DNA strands are separated. Within cells DNA occurs as chromosomes. The expression of genes is influenced by how the DNA is packaged in chromosomes.

In a DNA double helix the nucleobase on one strand only bonds with one type of nucleobase on the other strand and this is called complementary base pairing. Purines form hydrogen bonds to pyrimidines with adenine only bonding to thymine in two hydrogen bonds and cytosine only bonding to guanine in three hydrogen bonds.

DNA generally forms linear chromosomes in eukaryotes which include present day plants and animals with membrane-bound organelles. In prokaryotes which are single celled organisms (archaea and bacteria) that lack a membrane-bound nucleus DNA forms circular chromosomes. The set of chromosomes in a cell are its genome. In the human genome about 3 billion base pairs of DNA are arranged into 46 chromosomes.

All DNA functions rely on interactions with proteins which can be non-specific or the protein can bind specifically to a single DNA sequence. Enzymes can also bind to DNA and include polymerases that copy the DNA base sequence in transcription and DNA replication. DNA contains the genetic information that allows all modern organisms to grow, function and reproduce.

In genetic engineering genes are transferred within and across species boundaries to produce improved organisms. New DNA can be inserted into the host genome by initially isolating and copying the target genetic material using molecular cloning methods to generate a DNA sequence, or by synthesising DNA which is inserted into the host organism. Genes may be removed or knocked out by using a nuclease. Gene

5.0 THE NEW WORLD ORDER

targeting is a technique that uses homologous recombination to change a gene and can be used to delete a gene, add a gene or introduce point mutations.

Genetic engineering uses techniques that remove heritable material or that introduce DNA prepared outside of the organism either directly into the host or into a cell that is fused or hybridised with the host. Plants and animals that have been changed through genetic engineering are termed genetically modified organisms (GMOs).

In genome editing DNA is inserted, replaced or removed from a genome using artificially engineered nucleases or molecular scissors. Natural genome editing occurs through viral or sub-viral agents. Gene therapy is the genetic engineering of humans by replacing defective genes with effective alternatives.

In biotechnology cloning includes processes that are used to create copies of DNA fragments (molecular cloning), cells (cell cloning) or organisms. Artificial cloning of organisms is called reproductive cloning. Artificial embryo splitting or embryo twinning creates monozygotic twins from a single embryo.

Modern cloning techniques involving nuclear transfer have been successfully completed on several animal species. Human cloning which is the creation of a genetically identical copy of a human has not yet been attempted. Two possible types of human cloning are therapeutic cloning and reproductive cloning. Therapeutic cloning would involve cloning cells from a human for use in medicine and transplants. This technique is currently being researched. Reproductive cloning would be used to make an entire, cloned human.

The future of the human race is largely dependent on genetic engineering and cloning being able to accelerate the evolution of the human race. This is necessary for humans to continue developing the science and technology required to cater with future cataclysmic physical events in the Universe and prevent extinction.

The current human population of 7.4 billion is not sustainable without depleting Earth's resources. Although most of this population is largely non-productive, it is still rampantly consumeristic. The number needs to be reduced to a maximum of 2 billion individuals as soon as possible. Proper, global birth control is initially required. In future the birth rate will be controlled in embryo farms. Food storage and the development of chemosynthetic food production is required for 2 billion people, in case of an extinction event.

Genetic engineering needs to concentrate on the improved modifications to human intelligence, physical endurance, adaptability and behaviour. Sociological change is required so that human focus is primarily aimed at indefinite survival. Advancements in science and technology are paramount for this to occur. Natural selection needs to be replaced with programmed evolution which concentrates on higher intelligence and physical adaptation for space travel. This will allow for an organised global society in which each individual has a useful role.

Once embryo farms are established programmed breeding of selective individuals will be possible and human numbers can be limited. The result will be an organised self-controlled society which will progress naturally without any bureaucratic intervention. Time and resources will then be properly directed for accelerated human advancement.

Cloning has distinct advantages for the human race. Therapeutic cloning which would involve cloning human cells for use in medicine and transplants needs to be urgently implemented. Reproductive cloning which

5.0 THE NEW WORLD ORDER

would involve making entire, cloned humans, would assist in controlling and improving the gene pool. The DNA of scientifically brilliant people should be cryogenically preserved in liquid nitrogen so that they can be cloned in the future. With the advancement of genetically improved individuals having greater intelligence, preserving the DNA of brilliant individuals for future cloning will probably become redundant.

Without genetic engineering and cloning, the probability of human extinction is high. By relying on natural selection and a burgeoning population the human race will be overwhelmed by natural, cataclysmic events in the future. Rapid, human, evolutionary advancement is required for the ability to develop new, life sustaining technology.

5.4 Society and Infrastructure

In the New World Order complete globalisation is essential for proper social organisation. Globalisation is the process of integration of economies, industries, markets and policy making around the world. National and regional economies, societies and cultures become integrated through the global network of trade, communication, immigration and transportation. Large scale globalisation began in the 19th century.

For a properly functioning society considerable changes are required to social structure, ethics and values. There are currently 7.4 billion people on Earth most of whom are non-productive. At present society is dysfunctional and human values are misguided. Vast quantities of resources are being wasted on a burgeoning population for the construction of transient infrastructure such as roads, bridges, buildings and electrical grids. These facilities will eventually need to be replaced at considerable expense.

AN INTRODUCTION TO THE NEW WORLD ORDER

In the New World Order individuals will be genetically programmed for their roles in life and will be harmoniously integrated into an orderly society. It will be unnecessary to have bureaucracies which include government, armed forces, police, lawyers, debt collectors and expensive court jesters. Money will become obsolete in an ordered society.

Although government will be redundant in the New World Order centralist control is required during the transition to complete globalisation. This organisation should include members from the armed forces, businesses and academic institutions of the most powerful countries in the World. Its role will be to reduce the World's population to an acceptable size, accelerate genetic programming and human cloning procedures, change the main values in human life to rapid advancements in science and technology for indefinite survival, restructure society so that resources and human endeavour are used to advance the human race and remove all bureaucracy and social hindrances which bog down current society.

At present the media can be described as the opium of people. It is puerile, retrogressive and is determined to maintain the masses in a state of mediocrity. In the New World Order the media need intelligent commentators, instructive programming and a positive outlook for the future of the human race.

The education system requires transformation in the New World Order. Emphasis needs to be placed on science and mathematics to facilitate technological advancements for the future of the human race. Scientists and technologists should be classified as the most important and eminent persons for their contributions to human advancement.

Money will be unnecessary in the New World Order because genetically programmed individuals will know

5.0 THE NEW WORLD ORDER

their roles in life and all work will be organised for the common good and for technological progress. Nothing is more important than the dignified, indefinite survival of the human race in an organised, balanced society on Earth and in space.

Society requires suitable infrastructure for the civilised habitation of a reasonable number of human beings around the World. Infrastructure refers to the facilities and systems which serve a region and include highways, streets, roads, bridges, tunnels, mass transit, airports, water resources, sewers, solid-waste treatment and disposal, electrical grids, telecommunications and hazardous waste management. Soft infrastructure includes schools, universities, hospitals and maintenance facilities. Also related are operating procedures, management practices and development polices for changing demand.

Current infrastructure expenditure is increasing exponentially to cope with an uncontrolled human population explosion. Resources are being rampantly consumed and the environment is being destroyed to cope with a burgeoning, largely non-productive population. Until universal birth control is implemented, increased wastage of resources and human effort will continue to expand.

The New World Order requires significant changes to human thinking and values for the development of a progressive society in which technological advancement will overcome Nature's hurdles as they are confronted.

5.5 Technological Advances Which Will Assist Human Evolution

The human race has reached the evolutionary stage where it can take control of its destiny. The emphasis of a future, globalised society needs to be oriented towards the development of genetically improved humans and technology which protects humans from extinction events.

5.5.1 Goals in Genetic Engineering

The explosive evolution of non-productive, rampantly consumeristic humans needs to be reversed by birth control to an acceptable, maximum global population of 2 billion. In the New World Order human embryo farms will produce the required numbers of humans for the controlled progression of human life on Earth and in space. Genetic programming will produce the types of embryos needed to develop and accelerate technological advancement.

Efforts need to be concentrated on three aspects of genetic engineering. Human intelligence needs to be rapidly advanced so that technological advancements for survival are developed and implemented. Research into the human brain is paramount. Genetic engineering needs to be used to advance the parts of the brain responsible for intelligence. Conjunctively, artificial intelligence needs to be actively researched so that in the New World Order robotics will be used to complete all menial tasks.

The human anatomy is not designed for space travel. Because the Earth will eventually be destroyed human survival depends on extraterrestrial settlement. Genetic engineering is needed to adapt the human body for prolonged space travel in a weightless environment.

In the past diseases have had a significant effect on human mortality. With a current large population generally living in close proximity, a new contagious disease could quickly spread. Emphasis needs to be placed on genetic engineering research into rapid development of cures for new diseases, in particular, viruses.

5.0 THE NEW WORLD ORDER

5.5.2 Protection of Humans from Cataclysmic Events

In the foreseeable future cataclysmic events which could cause human extinction on Earth include a supervolcano eruption, collision with a large meteor or comet, or exposure to an intense gamma-ray burst. In about 5 billion years time the Sun will become a red giant engulfing Earth and obliterating all life.

In The New World Order, with a controlled global population of 2 billion, considerable effort and expenditure will be required to ensure human survival following a cataclysmic event. Resulting changes on Earth could include an extended volcanic winter, atmospheric pollution, world failure of food crops, death of livestock and damage to the ozone layer.

As well as reducing the human population to a controllable level which would prevent mass starvation, survival infrastructure and food reserves are required. Extensive underground survival accommodation is needed as well as preserved food supply including fungal spores which grow in dark conditions. Technology required to develop chemosynthetic protein needs to be urgently developed. It would take considerable effort to sustain a population of 2 billion on preserved and chemosynthetic food during an extended volcanic winter.

Research is required to determine if the largest possible hydrogen bombs mounted on rockets will disintegrate large meteors and comets prior to impact with Earth. Such bombs could be tested by shooting them at the Moon and measuring the sizes of the blast craters.

5.5.3 Future Trends in Science, Engineering, Technology and Recreation

In the New World Order major emphasis will need to be placed on science and technology for the accelerated evolution of the human race and for new developments to overcome the destructive forces of nature. For indefinite human survival, efficient space travel is also a top priority.

As well as requiring significant advances in human genetic engineering the education system needs to be streamlined to cope with rapidly evolving intelligence. A diversified education system is required for a range of individuals with variable genetic abilities. The main role of education will be to rapidly train individuals for their roles in a technical society. This will include the intelligentsia as well as those who are programmed for more labour intensive work.

The practice of science education will initially involve the learning of scientific principles. For intellectually programmed individuals progress to scientific research would then follow. The key items in the science curriculum will be chemistry, physics, mathematics, astronomy and genetic engineering. Once scientific principles are clearly understood they need to be practically applied to engineering and technology.

Science can be defined as the reasoned study of natural phenomena within the Universe with the aim of discovering enduring principles that build and organise knowledge in the form of testable explanations and predictions. Engineering is the application of science and practical knowledge to design, invent, build, maintain and improve structures, machines, tools, systems, components, materials and processes, by exploiting natural phenomena for human use. Technology is the collection of skills, techniques, methods and processes which are used in the production of goods and services.

In the New World Order the interrelationship of science, engineering and technology is essential for rapid advancements. There is huge scope for innovation and

5.0 THE NEW WORLD ORDER

new discoveries in these fields. In science there is still much to discover about matter, anti-matter, dark matter, forces, energy, quantum mechanics, teleportation, the Universe and genetic engineering. Together with engineering and technology exciting future fields include rapid advances in human genetic engineering, the use of chemosynthesis for food production, manipulation of climate and atmospheric conditions, improved solar energy utilisation, nuclear fusion, use of superconductors at room temperature, nanotechnology applications, teleportation and space travel research. Space travel which is in its infancy, is essential for indefinite human survival. Considerable research and development are required to attain high speed space travel and for the human body to adapt to weightlessness in space. Travel during suspended animation is a possibility.

Entertainment in the New World Order will change as the human brain and body evolves. Initially recreation will probably include sport, music, sex, drug stimulation, electro-therapy and the media. With increased intelligence, intellectual stimulation would become more important.

5.5.4 *Agriculture and Chemosynthesis*

In the New World Order food production will be required for everyday consumption and large storages are necessary for catastrophic events such as volcanic winters. Agriculture, which is the cultivation of animals, plants, fungi and other forms of life for food, fibre, fuel, medicines and other products to sustain human life, will remain important. Modern agronomy, plant breeding agrochemicals such as herbicides, pesticides and fertilisers and technological advancements have significantly increased yields from cultivation. Selective breeding and improved animal husbandry have increased meat production. In recent times features of agriculture include increased production, use of artificial fertilisers, herbicides and pesticides, selective breeding, and the introduction of genetic modifications to crops and animals. Disease and land degradation are two problems that require attention.

The main crops currently grown are cereals, pseudocereals, legumes, fruits, vegetables and fungi. Animal husbandry includes cattle, sheep, goats, pigs, buffalo, camels and poultry. Livestock production systems which are defined by feed source include grassland based, mixed and landless (feedlot). Until protein synthesis is fully developed, increased feedlot farming would reduce the stress of land degradation.

Significant improvements can be made to agriculture through genetic engineering. As many crops as possible should be made leguminous with nitrogen fixing bacteria nodules for increase ground fertility. Edible weeds need to be cultivated in the oceans. Fungi and similar plants should be able to flourish in the event of a volcanic winter. Protein synthesis needs to be developed to replace the use of livestock protein for human consumption.

Changes to the Earth's atmosphere which would adversely affect agricultural production include atmospheric contamination by volcanic ash, or dust from a large meteor impact resulting in a volcanic winter, increase in carbon dioxide level resulting in higher temperatures which would melt the polar ice caps and cause extensive flooding of low lands and damage to the ozone layer caused by a gamma-ray burst.

Accelerated atmospheric dust decontamination will require increased moisture to condense around dust particles to form cleansing rain droplets. A higher temperature is required to evaporate oceanic water and raise atmospheric humidity. The combination of

5.0 THE NEW WORLD ORDER

increased temperature and abundant atmospheric dust would facilitate rainfall and atmospheric cleansing.

The atmospheric carbon dioxide level is currently rising due to human activities, causing an increase in atmospheric temperature. In the future methods will be available to reduce the carbon dioxide concentration. In the middle Eocene epoch the drawdown of carbon dioxide occurred naturally when blooms of the freshwater fern Azolla flourished in the Arctic Ocean. The result was a change from a greenhouse to an icehouse climate which has remained. In future, the modern fern Azolla filiculoides could be used to reduce atmospheric carbon dioxide levels. Carbon sequestration is the process of carbon capture and the storage of atmospheric carbon dioxide, which counteracts global warming. It includes biological processes such as reafforestation, peat production, agricultural practices and seaweed blooming. Chemical processes such as mineral carbonation and limestone acid neutralisation also reduce carbon dioxide level.

Carbon dioxide can be injected into depleted oil and gas reservoirs and other geological structures such as coal mining goaves. Underground coal gasification is a process which converts coal into product gases including methane, hydrogen, carbon monoxide and carbon dioxide. The gasification occurs by injecting oxidants and returning product gas through production wells drilled from the surface. This process has significant potential but should be completed in isolation at depth, for containment. The goaves formed after gas release can be used to store carbon dioxide. This stored gas can later be used to feed carbon dioxide inhaling bacteria in ponds or large storages. Through genetic engineering, carbon dioxide inhaling bacteria can be used to produce ethylene plastic, branched alcohols for fuel and protein for use in feedlots and human consumption. Large scale production would assist in reducing the atmospheric carbon dioxide level.

Ocean related carbon sequestration has potential, if required. Ocean iron fertilisation can encourage phytoplankton growth with associated removal of atmospheric carbon dioxide. Ocean area fertilisation also encourages phytoplankton growth. Carbon dioxide could also be injected into deep-sea formations. Suitable injection sites need to be located for ocean floor sequestration.

It is possible for a gamma-ray burst to severely damage the ozone layer which occurs mainly in the lower part of the stratosphere. The ozone shield protects Earth by absorbing most of the Sun's ultraviolet radiation. It only has an ozone concentration of less than 10 parts per million. Recent damage to the ozone layer has occurred by the human use of chlorofluorocarbons. Research is required to determine how ozone can be injected into the stratosphere to remediate any future damaged ozone layer, particularly above human concentrations.

Chemosynthesis is the use of energy released by inorganic chemical reactions to produce food. Photosynthesis requires photosynthetic organisms containing chlorophyll to use solar energy to convert carbon dioxide and water into sugar and oxygen. Chemosynthesis uses hydrogen sulphide (or methane) and carbon dioxide to produce sugar, water and globules of sulphur. Chemosynthetic microorganisms (chemoautotrophs) include bacteria and archaea which inhabit dark regions of the oceans. Efficient food production by chemosynthesis requires considerable biochemistry and genetic engineering research to develop productive techniques for food supply on Earth, during space travel and on extraterrestrial colonies.

5.0 THE NEW WORLD ORDER

5.5.5 Solar Energy

Solar energy has huge potential for future human usage. It is light and heat from the Sun which are currently used in technologies such as solar heating, photovoltaics, solar thermal energy and artificial photosynthesis. The large magnitude of solar energy needs to be harnessed for widespread electricity production.

The Earth's surface receives about 122 000 terawatts of incoming solar radiation. Total solar energy absorbed by the atmosphere, occurs and land masses is about 3 850 000 exajoules, of which photosynthesis capture about 3000 exajoules in biomass. The untapped resource is immense.

Solar energy has a number of current applications. Solar hot water systems, which are efficient at geographical latitudes of less than 40 degrees, are used to heat water. Solar heating, cooling and ventilation technologies are now available for use in buildings. Solar cookers use sunlight for cooking, drying and pasteurisation and solar concentrators such as parabolic dishes and reflectors provide process heat for commercial and industrial use. Solar distillation is used to produce potable water and solar water disinfection sterilises water placed in plastic, polyethylene terephthalate bottles and exposed to sunlight for several hours. Solar power is produced by converting sunlight into electricity. This can be done directly using photovoltaics or indirectly by concentrating solar power using lenses or mirrors and tracking systems to form small beams which are converted into electric current using the photoelectric effect. Agriculture greenhouses convert solar light to heat which allows all year production.

Future solar research needs to focus on concentrating solar power for more efficient utilisation and improved conversion of solar energy to electricity. Technological improvements will include concentrated solar power tower plants, improved parabolic dishes, reflectors and solar magnifiers and more efficient photovoltaic converters.

The main problem with solar energy is that it is only available during daylight hours. Efficient techniques need to be developed to store converted solar energy during daylight hours for use at night. Current technology allows photovoltaic electricity to charge lithium batteries for night time usage. In future this method could be vastly improved. Pumped-storage hydroelectric schemes could be constructed to use solar energy to pump water uphill for storage and then produce hydroelectricity by downhill flow at night. Solar energy can also be stored at high temperatures by using molten salts which have a high specific heat capacity and are able to deliver heat at temperatures compatible with conventional power systems.

5.5.6   Nuclear Fusion

In nuclear fission a heavy atomic nucleus such as that of uranium or plutonium is subdivided into two fragments of about equal mass. During this process a large amount of energy is released. A nuclear reactor is a device which is used to start and control a sustained nuclear fission chain reaction. Nuclear reactors are used at nuclear power plants for electricity generation and for propulsion of ships. Heat from nuclear fission is used to produce steam which propels turbines. Nuclear reactors convert the energy released by controlled nuclear fission into thermal energy for further conversion to power or electricity. The power output of the reactor is adjusted by controlling the number of neutrons which are able to cause fission by the use of control rods. The two main problems with nuclear fission reactors is that they produce a considerable amount of toxic radioactive waste and in the past accidents at reactor sites have caused significant environmental damage.

5.0 THE NEW WORLD ORDER

The development of controlled nuclear fusion reactors will result in a cleaner and safer large source of future power. In nuclear fusion two or more atomic nuclei approach each other and then collide at high speed to form a new nucleus. The fusion of two nuclei with masses lower than iron -56 generally releases energy while the fusion of nuclei heavier than iron absorbs energy. The result is that lighter elements like hydrogen and helium undergo fusion whereas heavier elements like uranium and plutonium are fissionable.

The cause of the energy released in the fusion of light elements is due to the interaction of two opposing forces which are the nuclear force which combines protons and neutrons together and the Coulomb force which causes protons to repel each other. Because the nuclear force is stronger than the Coulomb force for smaller atomic nuclei fusion releases extra energy from the net attraction of these particles. Fusion reactions of the light elements power the stars and produce all elements in the process of nucleosynthesis.

Fusion reactions combine lighter atoms such as hydrogen together to form larger ones. The reactions take place at such high temperatures that the atoms have been ionised, with their electrons stripped off by heat to form a plasma. From an energy perspective the best fusion fuel is a one to one mixture of deuterium and tritium, which are hydrogen heavy isotopes. This mix has a lower Coulomb barrier or fusion barrier energy because of the high ratio of neutrons to protons. For practical fusion power systems temperature must be greatly increased to tens of millions of degrees and/or compressed to immense pressures. The end products of the fusion process are helium and water.

A number of fusion systems are currently at an early stage of research and considerable effort is required to develop economic fusion energy. Thermonuclear or hot fusion is currently used in thermonuclear weapons and produces extreme, thermal, kinetic energy due to collisions, which cannot currently be controlled.

Inertial confinement fusion or fusion by laser starts nuclear fusion reactions by heating and compressing a target pellet consisting of mixed deuterium and tritium. The target is hit by high-energy laser beams which explode the outer layer causing shock waves to travel inward through the target and cause nuclear fusion at the centre. The aim is to produce ignition of a significant quantity of the fuel. This technology requires new machines with larger ignition energies to approach success.

Magnetic confinement fusion or fusion in a magnetic field uses magnetic fields to confine fusion plasma. Early research used magnetic mirrors at each end of a solenoid. Toroidal machines include stellarators (torus configurations) and tokamaks (spheromaks with inserted central rods). To date spherical tokamaks with low toroidal fields are impractical for fusion neutron devices.

Tabletop fusion or sonoluminescence produces nuclear fusion at room temperature by bubble fusion which is the application of the rapid collapse of bubbles caused by sound waves or by pyroelectric fusion (crystal fusion), which is induced by an application of intense electric fields within a crystal. A tabletop particle accelerator has been developed that uses two opposing pyroelectric crystals to create a strong electric field. The device is filled with deuterium gas and the electric field removes electrons from the gas creating ions which are accelerated into a deuterium target on one of the crystals. When the target is struck neutrons are emitted indicating that nuclear fusion has occurred.

The mechanism of how sonoluminescence works is unknown. However during bubble collapse the inertia of the surrounding water causes high pressure and high temperature reaching around 10 000 degrees Celsius in the interior of the bubble causing some ionisation of any present noble gas. Single bubble sonoluminescence has the advantage of emitting more light due to less interactions between neighbouring bubbles. Multi-bubble sonoluminescence produces many oscillating and collapsing bubbles with less light emission. Tabletop fusion is at a very early stage of development with considerable research and development required for practical progress.

Beam-beam or beam-target fusion accelerator based light-ion fusion is a method using particle accelerators to obtain particle kinetic energies capable of inducing light-ion fusion reactions. If the energy required to initiate the reaction comes from accelerating one of the nuclei the process is called beam-target fusion but if both nuclei are accelerated it is called beam-beam fusion. Currently sealed-tube neutron generators are produced. These devices are small particle accelerators filled with deuterium and tritium gas where accelerated nuclei allow fusion to occur. Neutron generators are now commercially available for use in measurement equipment.

Antimatter generated fusion would require small amounts of antimatter to start a tiny fusion explosion. This technology is well beyond current practical possibility.

In 2015 the Max Planck Institute for Plasma Physics in Germany activated a stellarator which successfully produced and sustained hydrogen plasma for the first time. In a stellarator the magnetic confinement is produced with a single coil system, without a longitudinal net current in the plasma. This allows stellarators to operate continuously whereas tokamaks without auxiliary facilities operate in pulsed mode. The

Max Planck Institute successfully produced super-hot blobs of helium plasma in the stellarator. The team was also able to very briefly heat hydrogen gas to 80 million degrees.

A tokamak is a toroidal chamber which uses a magnetic field to confine plasma in the shape of a torus. For stable plasma equilibrium the magnetic field needs to move around the torus in a helical shape which is achieved by adding a toroidal field which is produced by electromagnets surrounding the torus and which travels around the torus in circles, and by a poloidal field which results from a toroidal electric current that flows inside the plasma and progresses in circles orthogonal to the toroidal field.

The required large toroidal currents cause the tokamak to have a stability problem which results in disruptions and damage. In present tokamak fusion experiments, insufficient fusion energy is produced to maintain the required plasma temperature. Despite a number of setbacks, controlled nuclear fusion is an exciting research field.

5.5.7 Superconductivity

Superconductors are materials that conduct electricity with no resistance and can carry a current indefinitely without losing any energy. No magnetic field can exist within a superconductor and the resistance abruptly drops to zero when the material is cooled below its critical temperature. A superconductor is classified as high temperature if it achieves a superconducting state when cooled using liquid nitrogen or is considered low temperature if increased cooling techniques are required to reach the critical temperature.

Conventional superconductors are rigid lattices of positive ions surrounded by electrons and electrical resistance occurs because electrons collide with the

5.0 THE NEW WORLD ORDER

lattices and lose energy. At low temperatures electrons can bond to each other to form Cooper pairs and at the same time the lattice becomes rigid enough to allow the coherent movement of waves called phonons. A phonon is a discrete quantum of vibrational, mechanical energy. Phonons and electrons are the two main types of elementary particles or excitations in solids.

Superconductivity occurs when the Cooper pairs and the phonons travel together through the material with the waves clearing the way for the electron pairs. It breaks down when the vibrations in the lattice, which is related to temperature, become strong enough to break apart the Cooper pairs. This is known as the critical temperature which until recently was about minus 230 degrees Celsius.

Superconductors have three distinct characteristics. Firstly, there is a sudden drop in electrical resistance when the material is cooled below the critical temperature. The second effect is that magnetic fields are expelled from inside the material and this phenomenon is known as the Meissner effect, which occurs during transition to superconductivity. The third characteristic is a change in the critical temperature when the atoms in the material are replaced with isotopes. The difference in the isotope mass causes the lattice to vibrate differently, which changes the critical temperature.

Ceramic substances discovered in the 1980's superconduct up to temperatures of about minus 110 degrees Celsius. In 2014 hydrogen sulphide was predicted to be a high temperature superconductor at a very high pressure. Current superconductors allow liquid nitrogen to be used as a refrigerant.

Superconductor magnets are very powerful and are used in magnetic resonance imaging and nuclear magnetic resonance machines, mass spectrometers and for beam-steering magnets in particle accelerators. Superconductor technology has significant future potential in applications such as superconducting generators in wind turbines, electric power transmission, transformers, power storage equipment, electric motors, magnetic levitation devices and magnetic refrigeration. Magnetic levitation or magnetic suspension is the method of suspending an object only using magnetic fields to counteract gravitational and other accelerations. Magnetic levitation requires a lifting force sufficient to counteract gravity and stability. By using superconductor magnets, levitation has enormous future potential for transport.

5.5.8 Nanotechnology

Nanotechnology is the engineering of functional systems at the molecular scale. It involves the manipulation of matter with at least one dimension sized from one to 100 nanometres. It is defined by size and includes fields of science such as organic chemistry, molecular biology, semiconductor technology, particle physics, surface science, microfabrication and development of new nanoscale materials. In the future nanotechnology will be able to create many new materials like nanomedicines, nanoelectronics, nanorobots, nanowire conductors, logic gates, data bits represented by the presence or absence of a single electron and low cost consumer products.

The emergence of nanotechnology occurred in about 1986 when K.E. Drexler published a book titled "Engines of Creation: The Coming Era of Nanotechnolgy." Since then commercialisation of products based on nanotechnology advancements has occurred.

There are two approaches in nanotechnology which are the "bottom-up" approach and the "top-down" approach. In the former materials and devices are constructed from molecular components. In the latter

5.0 THE NEW WORLD ORDER

nano-products are made from larger objects without control at the atomic level.

Nanotechnology is currently being actively researched with considerable scope for future innovation. Colloidal and interface science has resulted in new materials such as fullerenes (pure carbon forms having at least 60 atoms and existing as spheres, ellipsoids, tubes and many other shapes) nanorods and nanomaterials with fast ion transfer. Nanopillars are now used in solar cells and semiconductor nanoparticles are also found in display technology, lighting and biological imaging. Nanomaterials are being used in biomedical science for tissue engineering, transfer of drugs and as biosensors.

Molecular scale electronics develop molecules with useful electronic properties which can be used as single molecule components. Synthetic chemical methods can be used to make synthetic molecular motors. Bionanotechnology applies biomolecules to configure viruses, lipid assemblies and organic compounds.

Molecular nanotechnology manipulates single molecules in finely controlled ways. In nanorobotics emphasis is on self-sufficient machines operating on a nanoscale. The use of nanorobots in medicine is being researched. Important modern developments which are progressing include improvements to the scanning tunnelling microscope, atomic force microscope and the scanning acoustic microscope. The tip of the scanning probe can be used to manipulate nanostructures resulting in positional assembly.

Nanolithography techniques which have been developed include optical lithography, X-ray lithography and electron beam lithography. In lithography material is reduced to nanoscale size. Other nanotechnological advancements include fabrication of nanotubes and nanowires for semiconductor utilisation, atomic layer deposition, molecular vapour deposition and molecular self-assembly techniques such as regulating di-block polymers.

In bottom-up techniques larger structures are built on an atom by atom or molecule by molecule basis through chemical synthesis, self-assembly and positional assembly. Other methods include dual polarisation interferometry, molecular beam epitaxy and the emerging field of spintronics.

The top-down approach utilises larger objects to manufacture nano-products. Scanning probe microscopy using atomic force microscopes and scanning tunnelling microscopes are used to characterise and synthesise nanomaterials.

Nanotechnology will allow the development of much more powerful computers and communication devices as well as many applications in medical science. Two important concepts which need to be progressed are positional assembly and self-replication.

5.5.9 Teleportation

Teleportation is the transfer of matter, energy or information from one location to another without traversing the physical space between them. For teleportation to occur matter needs to be quantised, transposed in a dimensionless medium and then reconstituted. In order to dematerialise matter for teleportation the binding energy of the atoms and the nuclei needs to be overcome. Matter would then turn into radiation. Research into this field is just beginning with considerable future potential for transfer beyond the quantum level.

Physicists have been able to exchange information between photons and atoms as long as they were right next to each other. Successful teleportation on atoms

5.0 THE NEW WORLD ORDER

have been performed for the first time. Qubits (quantum bits) have been teleported from an atom to another with the help of a third auxiliary atom. This is achieved by entanglement whereby two particles can have related properties even when they are far apart. Quantum entanglement is the phenomenon that occurs when pairs or groups of particles are generated in such a way that the quantum state of each particle cannot be described independently but must be described for the system as a whole.

Teleportation has been achieved using different techniques but following the same basic methodology. Firstly a pair of highly entangled charged atoms or ions known as B and C are made. Next the state to be teleported is created in a third ion known as A. Then the B ion of the pair is entangled with the A ion. The internal state of both of these is then measured and the result is sent to ion C. This transforms the quantum state of ion C into that created for ion A, destroying the original quantum state of A. The teleportation successfully occurred in milliseconds. Fidelity is a measure of how well the quantum state of the second ion after teleportation resembles the original quantum state.

Quantum teleportation is the process by which quantum information can be transmitted from one location to another using classical communication and previously shared quantum entanglement between the sending and receiving locations. Maximum possible communication rate is the speed of light. Quantum teleportation is a form of communication which provides a way of transporting a qubit from one place to another without moving a physical particle along with it. Physicists have teleported the qubits encoded in the electronic state of atoms but they have not teleported the nuclear state or the nucleus itself. The distance of quantum teleportation has progressively increased and is now 143 km.

Teleportation technology has progressed rapidly since 1993. Physicists are currently attempting to develop a quantum computer in which could lead to the first fully realistic simulations of quantum phenomena.

The possibility of future human teleportation is currently beyond technological consideration. Extensive research and development are needed to initially come to terms with quantum entanglement, transportation and reconstitution on a scale greater than the atomic level. Teleportation possibilities could then be developed.

5.5.10 Space Travel and Extraterrestrial Colonisation

5.5.10.1 Introduction

Without space travel and extraterrestrial colonisation the human race cannot survive. Preliminary targets for human settlement should include the Moon and Mars. In future colonisation sites could include Ceres which is the largest object in the Asteroid Belt between Mars and Jupiter, Ganymede, Europa and Io which are moons of Jupiter and Titan and Enceladus, which are moons of Saturn.

The presence of water is essential for human survival. By electrolysis it can be converted to oxygen and hydrogen. Food production by chemosynthesis would be required in an extraterrestrial environment.

5.5.10.2 Space Travel

Considerable effort is required to advance space travel technology to a stage where humans can freely move within the Solar System and beyond. Much of the current expenditure which is being wasted on useless earthly infrastructure needs to be redirected to space travel research and extraterrestrial colonisation.

5.0 THE NEW WORLD ORDER

Advances are slow and are oblivious to the real possibility of future, globally catastrophic events which could affect human survival on Earth. Research needs to be focussed on much faster travel at greatly reduced cost. A number of options are being investigated.

Skylon is a British design for a reusable spaceplane that enters orbit as a single stage. It would be a hydrogen fuelled aircraft that would take off from a runway using atmospheric oxygen before switching the engines to internal liquid oxygen. The vehicle would re-enter the atmosphere being protected by a ceramic composite skin and would land on a runway.

Skylon is to be a fully reusable single stage-to-orbit vehicle. The engine design aims to be a good jet engine within the atmosphere as well as being an efficient rocket engine in space. The fuselage of Skylon is expected to be a carbon fibre reinforced polymer space frame that supports the weight of aluminium fuel tanks and to which a ceramic outer layer is attached.

The use of liquid hydrogen as a light and extremely powerful rocket propellant has been one of NASA's most significant technical achievements. Oxygen and hydrogen are gases that can only be liquefied at very low temperatures. Today liquid hydrogen is the signature fuel for space programs around the World. It was one of the significant technical achievements of twentieth century rocketry.

Nuclear powered spaceships have considerable future potential. The Voyager space probe currently heading beyond the Solar System and the Cassini spacecraft which is in orbit around Saturn, are fitted with nuclear power plants. Their radioisotope, thermoelectric generators depend on the natural decay of plutonium to generate heat, which is then converted into electricity. Future spaceships will need much more powerful propulsion systems than chemical rockets or solar powered probes. Nuclear systems such as fission or fusion would result in much more efficient power generation.

A fusion rocket driven by fusion power is currently a theoretical design which could provide efficient, ongoing acceleration in space without requiring large fuel reserves. The main advantage of fusion would be the very high specific impulse but the main disadvantage would probably be the large mass of the reactor. An electric generator operated by fusion power could be used to operate a spaceship. Another possibility is to direct the exhaust of fusion product out of the back of the rocket to provide the required thrust. Helium-3 fusion is a possible method for spacecraft propulsion. It is an isotope of helium with two protons and one neutron and could be fused with deuterium in a reactor to form a propellant. Helium-3 could be a power source because it occurs abundantly on the Moon due to concentration by solar wind colliding with the lunar surface.

For nuclear fusion to occur the plasma needs to remain confined. This can be done by magnetic confinement or by inertial confinement fusion. Magnetised target fusion is a new approach that combines the best features of magnetic confinement fusion with good energy confinement and inertial confinement fusion where there is efficient compression heating and wall free plasma containment. Magnetised target fusion uses plasma guns (electromagnetic acceleration, techniques) instead of powerful lasers resulting in cheaper and lower weight reactors. Another confinement concept for fusion rockets is inertial electrostatic confinement. A future concept worthy of consideration is antimatter catalysed nuclear pulse propulsion where tiny quantities of antimatter are used for fusion reactions.

The Bussard ramjet is a ramjet variant of a fusion rocket which would be capable of interstellar travel using huge

5.0 THE NEW WORLD ORDER

electromagnetic fields to thousands of kilometres in diameter, to collect and compress hydrogen from the interstellar medium. The high speeds force the reactive mass into a progressively constricted magnetic field with continued compression until thermonuclear fusion occurs. The main problem with this proposal is that the region surrounding the Sun has a much lower density of interstellar hydrogen than was previously believed. An auxiliary nuclear fuel supply is required for the system to work.

Ion propulsion is being used to keep telecommunications satellites in their required geosynchronous orbits and to propel spacecraft through the Solar System. An ion thruster ionises propellant by adding or removing electrons to produce ions. Most thrusters ionise propellant by electron bombardment. A high energy electron with a negative charge collides with a propellant atom of neutral charge, releasing electrons from the propellant atom which results in a positively charged ion. The gas produced consists of positive ions and negative electrons in electrically neutral proportions. The resultant plasma which is similar to lightning and the substance inside fluorescent light bulbs, has the properties of gas and is affected by electric and magnetic fields. Xenon is the most common propellant used in ion propulsion because it has a high atomic mass and is capable of generating the required level of thrust when ions are accelerated. It is inert and has a high storage density. In most thrusters electrons are generated through the discharge hollow cathode by a process known as thermionic emission.

Electrons produced by the discharge cathode are attracted to the discharge chamber walls which are charged to a high, positive potential by the voltage from the thruster's discharge power supply. A neutral propellant is injected into the discharge chamber, where electrons bombard the propellant to produce positively charged ions and more electrons. Strong magnets prevent the electrons from reaching the discharge chamber walls which increases the probability of an ionising event. The positively charged ions migrate toward grids containing thousands of precisely aligned apertures at the rear end of the ion thruster. The first grid is a very highly positively charged electrode that is configured to force the discharge plasma to exist at a high voltage. As ions pass between the grids they are accelerated towards the accelerator grid which is a negatively charged electrode. The positively charged ions are accelerated out of the thruster as an ion beam which produces thrust. The neutraliser is another hollow cathode which expels an equal amount of electrons to make the total charge of the exhaust beam neutral.

The main components of an ion propulsion system are the ion thruster, power processing unit, propellant management system and the digital control and interface unit. The power processing unit converts the electrical power from the solar cells or nuclear heat power sources into the voltages required for the hollow cathodes to operate, to polarise the grids and to provide the currents needed to produce the ion beam. The propellant management system controls the xenon pressure for accurate ion thrusting.

Current ion propulsion is being used to keep over 100 geosynchronous Earth orbit communication satellites in their required locations. Ion thrusters are now allowing spacecraft to travel deep into the Solar System. In future new thruster design will achieve very high power and thrust levels resulting in reduced system complexity and enhanced performance.

Research is progressing to develop a space shuttle which uses microwave energy delivered wirelessly to a heat exchanger on the vehicle. Instead of chemical combustion the spaceplane will be propelled by ejecting hydrogen, heated with the microwave energy as it flows

5.0 THE NEW WORLD ORDER

through the heat exchanger and will be exhausted through the nozzle, creating thrust. A phased array of microwave antennas located on the ground would deliver the microwave energy during the entire ascent into space. The externally powered launch vehicles will be a significant improvement in efficiency compared to the chemical rockets which are currently being used. Microwave technology will result in single stage-to-orbit spaceplanes and aircraft type operations to orbit at significantly reduced costs.

Solar sails which are also known as light sails or photon sails, are a proposed method of spacecraft propulsion which would use large mirrors driven by radiation pressure caused by sunlight. Photons cause the movement in the vacuum of space. To travel to the outer Solar System solar sails would orbit close to the Sun and then fly outwards at speeds to 750 000 km per hour.

The force exerted on a solar sail and the acceleration of the craft vary by the inverse square of the distance from the Sun and by the square of the cosine of the angle between the sail force vector and the radial from the Sun. If some of the energy is absorbed, this energy would heat the sail which re-radiates that energy from the front and rear surfaces. The solar wind which is the flux of charged particles blown from the Sun exerts a nominal dynamic pressure of about 3 to 4 nPa, which is about three orders of magnitude less than the solar radiation pressure on a reflective sail.

Sail loading which is the total mass divided by the sail area, is an important parameter. An active attitude control system is needed for a sail craft to achieve and maintain the desired orientation. The required sail orientation changes slowly in interplanetary space but much more rapidly in a planetary orbit. Attitude control could be achieved with control vanes movement of individual sails, movement of a control mass or by altering reflectivity.

The physical environment for sail craft utilisation ranges throughout the Solar System and would vary from near the Sun to beyond Neptune. Such craft could economically travel to and from all of the inner planets. Travel to the outer planets would involve solar swing-by to increase speed. For interstellar flight propulsion schemes using lasers or masers could be used to push giant sails to a significant fraction of the speed of light. The engineering challenges to develop such technology are enormous.

Current designs of solar sails consist of a thin layer of aluminium coating on a plastic sheet. A high performance sail could be made of lithium. Magnesium and beryllium are other options. Sails are currently fabricated on Earth. They are packed, launched and unfurled in space. In the future construction could be completed in orbit. Swing-by manoeuvres using the Moon or the Sun would increase a craft's energy.

The first interplanetary solar sail spacecraft was launched in 2010. Manned space flight using solar sails requires considerable technological development. A current project is designed to develop a sail craft which would travel to Alpha Centauri at 20% the speed of light.

A wormhole is a hypothetical topological feature which is a shortcut connecting two separate points in space. It is a tunnel with two ends which connect separate points. Traversable wormholes would allow travel in both directions from one part of the Universe to another part. If traversable wormholes exist they could alter the speed of time and allow time travel. The possibility of wormhole travel is currently an imaginative concept.

5.0 THE NEW WORLD ORDER

5.5.10.3 Human Adaptation to Space Travel

Human physiological evolution has been significantly influenced by Earth's gravity. Space travel requires considerable adaptation to a weightless environment. On return to Earth the human body needs to readjust to gravitational effects. A number of techniques need to be investigated to ensure that physiological damage does not occur due to prolonged weightlessness followed by exposure to Earth's gravity.

5.5.10.3.1 Vacuum Suit

A vacuum suit is a form or protective suit used for extra-vehicular activity in outer space. It is used to protect the wearer from the effects of space vacuum, prevent asphyxiation and provide protection from dust and other particles. Vacuum suits are completely sealed and have their own oxygen supply. They have articulated joints to allow the wearer to complete tasks in space.

Various models of vacuum suits have been developed for different applications. Space suits are currently used for extra-vehicular activities in Earth's orbit.

5.5.10.3.2 Vacuum Chamber

A vacuum chamber is a rigid enclosure from which air and other gases have been removed by a vacuum pump to form a low pressure environment. Such chambers allow experiments to be done on mechanical devices which need to operate in outer space. A thermal vacuum chamber provides a thermal environment which simulates conditions in space.

Chambers often have multiple ports covered with vacuum flanges. This allows instruments or windows to be installed in the walls of the chamber.

Experimentation is required to determine the physiological effects of prolonged human exposure to a vacuum environment. Results could determine if there are any physical advantages for extended, human space travel.

5.5.10.3.3 Artificial Gravity

Artificial gravity is an acceleration caused by the application of a force. It is an important issue in reducing adverse health effects caused by prolonged weightlessness during manned space flight. Without gravity there is significant muscle and bone loss in animals.

Artificial gravity can be generated in various ways. A rotating spacecraft would drive any inside object towards the hull by centrifugal force. The gravity felt by the object is the reaction force of the object on the hull in response to the centripetal force of the hull on the object. Artificial gravity levels vary proportionately with the distance from the centre of rotation. The engineering requirements for having a rotating spacecraft with artificial gravity are technically feasible.

Linear acceleration can provide significant g-force. A spacecraft having constant, linear acceleration would appear to have a gravitational pull in the direction opposite to the acceleration. Current rockets can only provide temporary acceleration due to the limited supply of rocket fuel.

A similar effect to gravity has been achieved through diamagnetism. Magnets with extremely powerful magnetic fields are required. This application is not currently practical due to expensive cryogenics required for superconductivity or the need for several megawatts of power.

5.0 THE NEW WORLD ORDER

A number of proposals are incorporating artificial gravity into their design by utilising a rotational gravity habitat for long duration travel. New human centrifuges are being used to explore the use of artificial gravity as a way of overcoming the detrimental effects of extended weightlessness.

5.5.10.3.4 Suspended Animation

Suspended animation or reversible metabolic hibernation is a state in which the body process (such as blood circulation, breathing and heart beat) stop or become very slow for a period of time, while the person or animal is unconscious. Organisms including embryos have been cryogenically preserved and revived after time periods to 13 years. Placing astronauts in suspended animation has been proposed as a method for interstellar or intergalactic travel.

To be able to achieve successful suspended animation a reliable method is needed to prevent damage to cells. Research projects are currently investigating how to achieve induced hibernation in humans. NASA is developing a hibernation chamber which is a cryosleep system where humans are chilled to artificially induce a state of hypothermia for hibernation periods to two weeks.

5.5.10.3.5 Genetic Engineering

Considerable research is required into the human genome to determine how humans can be genetically modified for extended space travel. Astronaut enhancement needs to include genetic modification to prevent loss of bone, muscle and strength during weightlessness, improved blood circulation and maintenance of normal metabolic functions.

Future genetic engineering will need to modify human physiology for extended space travel and life in very different environments.

5.5.10.4 Extraterrestrial Colonisation

Extraterrestrial colonisation is permanent human habitation away from planet Earth. The main reason for space colonisation is to ensure the indefinite survival of the human race in the event of natural or man-made disasters.

Materials and energy resources in space are enormous. Because colonisation costs are currently prohibitive considerable research and development are needed to develop efficient transport and survival techniques. Two possible types of space colonies are surface based settlements on or below the surfaces of planets or moons and free floating stations known as space habitats. The former has the advantage of being more permanent whereas the latter type eventually crashes due to gravitational attraction.

The easiest way to build colonies in space is to tunnel into local rock. Efficient tunnelling machines are required to construct subterranean habitats. Access would be required to water, oxygen, food, construction materials and energy. Local water supply would be essential to be used and for oxygen generation by electrolysis. Food could be produced by chemosynthesis. Nearby materials would need to be extracted for construction. Solar energy in orbit is abundant and reliable. Large solar powered photovoltaic cell arrays or thermal power plants would be required to generate electricity. Radiation protection would naturally occur in subterranean habitats.

The Earth's Moon is the logical, initial target for a permanent space station and colonisation. It is close to Earth and has a low escape velocity. On the lunar surface

5.0 THE NEW WORLD ORDER

the requirements for first settlement are proximity to rock outcrops where tunnel portals can be excavated, adjacent water supply and ready access to materials such as metals, silicon and oxygen.

Mars would be the second target for colonisation. The Martian atmosphere is very thin due to the lack of a magnetic field. Radiation is intense on the surface and dust storms are common. A settlement site would require proximity to rock outcrops for tunnelling and an adjacent water supply. Carbon dioxide is present in the atmosphere.

The terraforming of Mars would be a process where the climate and surface would be changed to make large areas of planet more hospitable for human habitation. The three main required changes are enhancing the magnetosphere, increasing the atmospheric gas content and raising the temperature. The atmosphere could be improved by importing ammonia, hydrocarbons including methane and fluorine compounds. Orbital mirrors could be used to increase the surface temperature and sublimate the carbon dioxide ice sheet. Martian experiments using photosynthetic algae, cyanobacteria and methanogens would be useful for future biodomes and atmospheric enhancement.

Protecting the Martian atmosphere would require creating an artificial magnetosphere which is not currently feasible. The overall energy required to sublimate the carbon dioxide from the south polar ice cap is currently prohibitive.

The largest body in the Asteroid Belt is Ceres which is a dwarf planet. It is strategically located for future asteroid mining activity. Ceres has water and it could be terraformed to be more hospitable for colonists.

Three of Jupiter's moons named Europa, Ganymede and Callisto are potential sites for future colonisation. The main problems with the Jovian system are the presence of a severe radiation environment and a particularly deep gravity well. Europa is considered to be one of the more habitable locations in the Solar System and an investigation for signs of life is warranted.

Ganymede is the largest moon in the Solar System and it has a magnetosphere which reduces surface radiation. The magnetosphere is probably due to the presence of a molten core.

Callisto is at a significant distance from Jupiter's powerful radiation belt. It could be possible to build a surface settlement base for further exploration of the Solar System.

Europa, Ganymede and Callisto have an abundance of volatiles which would facilitate future colonisation.

The moons of Saturn which are possible colonisation targets include Titan and Enceladus. Titan is the only moon in the Solar System to have a dense atmosphere and is rich in carbon compounds. The surface consists of water ice with lakes of liquid hydrocarbons in the polar region. Titan is probably the most hospitable extraterrestrial location in the Solar System for human colonisation.

Enceladus is a small, icy moon which orbits close to Saturn. Plumes of ice and water vapour erupt from the southern polar region. Liquid water appears to occur below the surface, which would be beneficial for colonisation.

The moons of Uranus and Neptune have large volumes of frozen water and other volatiles. These areas are very cold with frozen nitrogen occurring on the surface.

The Trans-Neptunian Region is that area beyond the orbit of Neptune. The Kuiper Belt is estimated to have

5.0 THE NEW WORLD ORDER

more than 70 000 bodies greater than 100 km in size. The Oort Cloud contains up to a trillion comets.

There are significant problems associated with colonisation outside the Solar System which include the considerable distance from Earth, extremely cold temperatures and the lack of solar energy for solar power. For space colonisation and intergalactic travel, high but sub-relativistic speeds would at least be required with vast energy reserves being transported. It could become feasible to utilise Alcubierre warp drives to achieve faster than light speed travel if a configurable energy density field lower than that of a vacuum could be created. The result would be negative mass. A spacecraft would not exceed the speed of light within a local reference frame, but would traverse distances by contracting space in front of it and expanding space behind it resulting in faster than light speed travel. Objects are unable to accelerate to light speed in normal space-time. Alcubierre drive would shift space around an object so that it would arrive at its destination faster than light would in normal space.

6.0 CONCLUSION

The human race has two choices which are either to adopt the New World Order and live in dignity forever or to continue aimlessly and suffer ignominious extinction. By accepting the New World Order the human race will become the Master of the Universe.

7.0 REFERENCES

1. The Standard Model of Elementary Particles, 2009. http://en.wikipedia.org/wiki/Standard_Model

2. National Geographic, Society, 2005. Encyclopedia of Space.

3. Extensive reference has been made to Wikipedia particularly in Section 4.0. https://en.wikipedia.org/

4. The images on the covers are from NASA.

www.ingramcontent.com/pod-product-compliance
Lightning Source LLC
Chambersburg PA
CBHW071813200526
45169CB00017B/206